# Poverty, Vulnerability, and Agricultural Extension
## Policy Reforms in a Globalizing World

# Poverty, Vulnerability, and Agricultural Extension

*Policy Reforms in a Globalizing World*

*Editors*

IAN CHRISTOPLOS
and
JOHN FARRINGTON

OXFORD
UNIVERSITY PRESS

# OXFORD
UNIVERSITY PRESS

YMCA Library Building, Jai Singh Road, New Delhi 110 001

Oxford University Press is a department of the University of Oxford. It furthers the
University's objective of excellence in research, scholarship, and education
by publishing worldwide in

Oxford New York
Auckland Bangkok Buenos Aires Cape Town Chennai
Dar es Salaam Delhi Hong Kong Istanbul Karachi Kolkata
Kuala Lumpur Madrid Melbourne Mexico City Mumbai Nairobi
S±ao Paulo Shanghai Singapore Taipei Tokyo Toronto

Oxford is a registered trade mark of Oxford University Press
in the UK and in certain other countries

Published in India
By Oxford University Press, New Delhi

© Oxford University Press 2004

The moral rights of the author have been asserted

Database right Oxford University Press (maker)

First published 2004

ISBN 0 19 566826 X

Typeset in Garamond in 10.5/12
by Excellent Laser Typesetters, Pitampura, Delhi 110 034
Printed at Sai Printopack Pvt. Ltd. Y-56, Okhla, Phase-I, New Delhi-110 020
and published by Manzar Khan, Oxford University Press,
YMCA Library Building, Jai Singh Road, New Delhi 110 001

# Acknowledgements

This volume is based on case studies funded by the UK Department for International Development (DFID) and the Swedish International Development Co-operation Agency (Sida), under the auspices of the Neuchâtel Initiative, an informal group of donors and policy analysts seeking to identify new approaches to agricultural extension. The editors are indebted to DFID and Sida, and to members of the Neuchâtel Initiative for comments on earlier versions of the case studies. The views presented here, however, are those of the case study authors and editors alone, and do not necessarily correspond with the official views of DFID, Sida, or any other agency.

# Contributors

MALIN BECKMAN
Lecturer, Department of Rural Development and Agroecology
Swedish University of Agricultural Sciences (SLU), Sweden

ALAN J. BOJANIC
Vice Minister
Ministerio de Argicultura y Ganaderia y Desarrollo Rural (MAGDR)
La Paz, Bolivia

IAN CHRISTOPLOS
Head of the Department of Rural Development and Agroecology
Swedish University of Agricultural Sciences (SLU), Sweden

JOHN FARRINGTON
Research Fellow, Overseas Development Institute
and Visiting Professor
International and Rural Development Department
Reading University Overseas Development Institute, London, UK

GEORGINA HOLT
Research Fellow
Department for International and Rural Development
University of Reading, UK

ANDREW D. KIDD
Sustainable Livelihoods Adviser
DFID (Nigeria), British High Commission, Nigeria

RASHEED SULAIMAN V.
Scientist, Institutional change
National Centre for Agricultural Economics and Policy Research
Pusa, New Delhi, INDIA

# Contents

# Tables and Figures

## TABLES

## FIGURES

# Boxes and Maps

## BOXES

MAPS

# Abbreviations

| | |
|---|---|
| ADP | agricultural development project |
| AEP | agricultural extension programme |
| ATAIN | agribusiness training and input network |
| ATIC | agricultural technology information centre |
| ATM | mass technical assistance |
| ATMA | Agricultural Technology Management Agency |
| CD | community development |
| CDF | comprehensive development framework |
| BARC | Bolivian agricultural research centre |
| CIAT | Centre for Tropical Agriculture |
| CLUSA | Co-operative League of the United States |
| CPRGS | comprehensive poverty reduction and growth strategy |
| CRS | Catholic Relief Services |
| CSS | centrally sponsored schemes |
| DoA | Department of Agriculture |
| DPI | Department of Planning and Investment |
| DTC | Department of Transfer and Communication |
| EGS | Employment Guarantee Scheme |
| EU | European Union |
| FAIDA | Finance and Advice in Development Assistance |
| FAO | Food and Agriculture Organization |
| FCI | Food Corporation of India |
| FCP | Vietnam–Sweden Forestry Co-operation Programme |
| FDTA | Foundation for Development of Agricultural Technology |

| | |
|---|---|
| FIAC | farm information and advisory centres |
| FPR&E | farmer participatory research and extension |
| GDP | Gross Domestic Product |
| GM | genetically modified |
| GoI | Government of India |
| HIPC | highly indebted poor countries |
| HIV/AIDS | human immunodeficiency virus/acquired immune deficiency syndrome |
| HRD | human resource development |
| IADP | intensive agricultural district programme |
| IBTA | Bolivian institute of agricultural technology |
| ICAR | Indian Council of Agricultural Research |
| ICDS | Integrated Child Development Services |
| ICT | information and communication technology |
| IDEA | investment in developing export agriculture |
| IDPs | internally displaced people |
| IMF | International Monetary Fund |
| INTA | National Institute for Agricultural Technology |
| IPM | integrated pest management |
| IRDP | Integrated Rural Development Programme |
| ISPs | internet service providers |
| IT | information technology |
| ITD | innovations in technology dissemination |
| IU | intermediate users |
| IVLP | institute village linkage programme |
| KVK | krishi vigyan kendra (farm science centre) |
| LRA | Labour Research Association |
| MAGFOR | Ministry of Agriculture, Livestock, and Forestry |
| MANAGE | Institute of Agricultural Extension Management |
| MARD | Ministry of Agriculture and Rural Development |
| MoA | Ministry of Agriculture |
| MOLISA | Ministry of Labour, Invalids, and Social Affairs |
| MRDP | Mountain Areas Rural Development Programme |
| MTN | Mobile Telephone Networks |
| NAADS | National Agricultural Advisory Services |
| NATP | National Agricultural Technology Project |
| NES | National Extension Service |
| NGO | non-governmental organization |
| NI | Neuchâtel Initiative |
| NPM | new public management |

| | |
|---|---|
| NRM | National Resistance Movement |
| NSI | National Systems of Innovation |
| OBC | other backward caste |
| OECD | Organization for Economic Co-operation and Development |
| OTBs | Territorial Based Organizations |
| PAF | Poverty Action Fund |
| PCAC | Farmer to Farmer Organization |
| PDS | Public Distribution System |
| PEAP | Poverty Eradication Action Plan |
| PMA | Plan for Modernization of Agriculture |
| POPs | points of presence |
| PPA | participatory poverty assessment |
| PPL | popular participation law |
| PROINPA | Foundation for Research and Promotion of Andean Crops |
| PRSP | Poverty Reduction Strategy Papers |
| PTD | participatory research and extension |
| SAI | Inter-american Agricultural Service |
| SAU | State Agricultural University |
| SC | scheduled caste |
| SDC | Swiss Agency for Development and Co-operation |
| SFAC | Small Farmers Agri-business Consortium |
| SHG | self-help group |
| Sida | Swedish International Development Co-operation Agency |
| SMS | subject matter specialist |
| SNV | Netherlands Volunteer Service |
| ST | scheduled tribe |
| T&V | training & visit |
| TAR | technology assessment and refinement |
| UNAG | National Union of Farmers and Ranchers |
| UNDP | United Nations Development Programme |
| USAID | US Agency for International Development |
| VCCA | Vietnam Coffee and Cocoa Association |
| VEW | village extension worker |
| VLPA | village level participatory approach |
| VPDO | village panchyat development officer |
| WTO | World Trade Organization |
| ZARS | zonal agricultural research stations |

# 1
# The Issues
*Ian Christoplos • John Farrington*

## BACKGROUND

This book brings together a set of studies on the relevance of agricultural extension to poverty and vulnerability in a context of globalization. The authors look at experiences in India, Vietnam, Uganda, Nicaragua, and Bolivia to raise questions regarding the potential for refocusing extension efforts so as to make them more pro-poor. The efficacy of current policy frameworks in different countries is analysed, and alternatives for the future are assessed.

The studies were conducted under the auspices of the Neuchâtel Initiative (NI), an informal group of donors and policy analysts aiming to identify new approaches to agricultural extension. The NI arose out of a consensus among development partners that agricultural extension systems should play a major role in economic and social development. Whilst the importance of extension was clear, nonetheless, there was an uncertainty about how these systems should operate. Controversies around extension alternatives involve institutional and conceptual, as well as financial issues. The NI has been organizing meetings to discuss these issues since the mid-1990s to explore where a more common vision can be attained regarding the place of extension in the future of rural development. The main features of the 'common vision' developed by NI are summarized in Box 1.1.

## PRO-POOR EXTENSION AND LIVELIHOODS

This introductory chapter highlights the basic questions and hypotheses that guided the individual country studies. The central assumption underpinning these studies is that pro-poor extension neither involves

---

BOX 1.1
NI Common Framework on Agricultural Extension

Driving forces for change
Several trends are exerting their influence on the current situation which makes reform essential:

- changes are afoot in many countries: decentralization, liberalization, privatization, and democratization;
- new actors are becoming involved in 'extension' activities;
- public spending on extension is shrinking;
- the aims of official development assistance are becoming more focused.

The NI vision of extension
In the light of the changes taking place, the common framework outlines six key principles of a vision for extension:

- a sound agricultural policy is indispensable;
- extension is 'facilitation', as much as if not more than 'technology transfer';
- producers are clients, sponsors and stakeholders, rather than beneficiaries of agricultural extension;
- market demand creates an impetus for a new relationship between farmers and private suppliers of goods and services;
- new perspectives are needed regarding public funding and private actors; and
- pluralism and decentralized activities require co-ordination and dialogue between actors.

Proposal for the engagement and co-ordination of donors
The common framework also recognizes six key avenues for the engagement and co-ordination of donors:

- support negotiated national policy making between actual stakeholders;
- consider the long-term financial viability of extension activities;
- include exit strategies in all planning;
- facilitate funding of producer initiatives;
- ensure that extension activities are flanked by support for agricultural training, farmer organizations, and agricultural research;
- establish closer co-ordination between co-operation agencies.

*Source:* Neuchâtel Group, 1999, *www.neuchâtelinitiative.net*

---

a simple return to past approaches of supporting subsistence farming and other activities often designated as 'food security', nor is it a matter of redoubling efforts to ensure that standard messages (which these days focus primarily on commercialization and market orientation) are better targeted at the poor. Pro-poor extension, and extension that takes vulnerability into account differ fundamentally from current practice. The essential starting point must shift from a desire for productivity increase to an analysis of who the poor are and why they are poor. Furthermore, a much broader view of poverty is needed. Lack of income is not the sole aspect of being poor; poverty is above all related to a lack of power and lack of entitlements. If one considers why the poor so rarely benefit directly from extension and other rural services, it directs attention to a view of poverty as a lack of entitlements to these services (Sen and Drèze 1989).

If poor farmers are unable to draw on such services, we must then ask why the entitlements of the poor continue to be limited, and how they themselves strive to use their assets to access information about changing agricultural technologies. Extension is about spreading new knowledge, and to understand the potential for pro-poor extension we must begin by looking at how the poor construct their own knowledge systems. Livelihoods approaches are a useful starting point in identifying the routes that poor people use to deal with their poverty. Livelihoods analysis sees the poor as having access to five types of 'capital assets' (physical, natural, financial, social, and human)[1] which they can draw upon selectively in the pursuit of livelihood activities to achieve desired outcomes. The priorities of the poor typically include increased income, reduced vulnerability, more reliable access to resources, and other aspects of well-being (Figure 1.1). The poor will typically pursue a 'portfolio' of activities, which they will change with shifting pressures and opportunities. Poverty in this sense is dynamic: people can move in and out of it, depending on individual, local, national, or international circumstances. The relationship between the activities that the poor choose to pursue, and the outcomes that they desire, will be affected by policies, institutions and processes, both formal and non-formal. For this reason, the studies in this volume look beyond specific extension methods and structures; to focus more on policy frameworks, institutional actors, and local and international processes that have an impact on how they choose their livelihood strategies and if these strategies succeed or fail.

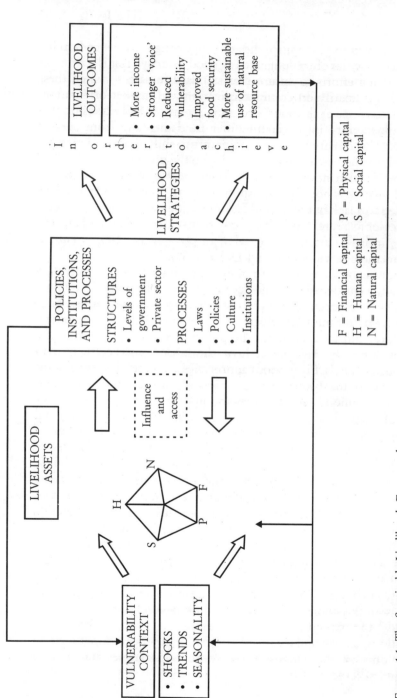

FIGURE 1.1: The Sustainable Livelihoods Framework

4

Of the wide range of policies, institutions, and processes that influence extension to be better related to the livelihood strategies of the poor, two merit particular mention.

The first is that the poor face substantial transaction costs in accessing their means of production, in adding value to their produce, and in accessing markets for it. A fundamental purpose of rural policy—of which the extension provision is a component—is to reduce these transaction costs. Poor people try to follow livelihood options using the assets that they possess and the pressures and opportunities around them. But they are 'fallible learners' (Ostrom *et al*. 1993), who make choices based on incomplete knowledge of alternatives and the likely outcomes. The core of pro-poor extension is to help people make better choices—through the supply of information and enhancement of their capacity to process information and to act on it—thereby reducing the transaction costs involved in productive activities or livelihood options.

A second is that in many countries the policies, institutions, and processes are loaded against the poor. Although some traditional institutions provide safety nets, there are indications that these are weakening in the face of globalization (Bryceson 2000). In other societies, traditional systems such as caste, may actually perpetuate social and economic difference. Prejudice against engagement of women in economic activity is, for example, widespread. Organizations in the private and public sectors are also less likely to serve the poor in difficult areas where there is little competition in input or produce markets, less supervision of junior civil servants, and, a higher number of vacant posts.

The studies in this book make it clear that extension cannot act in isolation if it is to find ways of addressing acute poverty. Safety nets become more important towards the lower end of the income or asset scale. Indeed, for those unable to engage in productive activity (such as women with a large numbers of dependents, young orphans, or the old and the chronically sick) safety nets provide practically the only means of support. One immediate implication is that in areas characterized by high levels of chronic illness such as human immunodeficiency virus (HIV)/acquired immune deficiency syndrome (AIDS) and high dependency ratios, agricultural policy (and with it, extension) must adapt to the types and levels of labour still available; however, it will, by itself, not be sufficient to address extreme poverty and must be accompanied by strengthened safety nets.

It is now well-known that technical change in agriculture[2] can impact on the poor in six possible ways:

1. directly through increases in on-farm production, whether consumed directly or marketed;

2. through increased local employment opportunities;

3. through secondary effects on growth in the local non-farm economy;

4. through increased migration opportunities that it provides;

5. through increased provision of staple food and fibre; and

6. through pro-poor price effects in all of the above (relatively lower food prices, higher wages, increased returns to own-farm labour, etc.)

These possible effects interact in different ways, and much will depend on the distribution and sequencing of technical change. For instance, lower food prices will almost universally benefit the poor, since all are consumers, but some are also producers. If the technical change underpinning reduced prices is first introduced by larger-scale farmers, then poorer ones may find themselves unable to compete in the changed markets. The extent of this problem will vary from country to country, depending on the prevailing levels of inequality. Regulatory and quality control institutions may play an increasing role in either supporting or preventing poor people's access to changing global markets. The poor may not have the resources to meet new and rapidly changing demands, or the risks may be too great. How technical change impacts on the poor is, therefore, dependent, to a high degree, on their poverty context or the 'rules of the game'. An important question is how technical change can best be managed or influenced to raise the probability that impacts on the poor will be positive.

A further complicating factor is that the livelihood strategies of the poor are dynamic, complex, and diverse: a single household (even a single individual) may engage in a range of own-account or labouring activities in farming or the non-farm economy. Complex and institutionally related factors, such as improved access to resources, reduced risk, and enhanced voice in planning rural 'futures', are also potentially important. Diverse livelihood options of poor people can also lead to a competition for scarce resources, for example, among farmers and pastoralists, leading to conflict. It may not always be possible to construct win–win strategies and 'whose voice counts?' becomes of great importance. Understanding of at least the broad and shifting patterns of rural livelihoods and the power relations influencing them is, therefore, a prerequisite to the introduction of pro-poor technical change, as also is the capacity to monitor impacts and make necessary course-corrections. In this sense, a process approach is necessary (Mosse *et al.* 1998). The studies in this volume describe the difficulties that mainstream extension services have had in dealing with such complexities in an open, client-oriented manner. Their failures have led to the growth of different sets of extension actors that are more attuned and more committed to engaging in the often politicized arenas where the voice of their clients must be heard.

## THE ROLES AND ORGANIZATION OF EXTENSION

There are always a number of factors involved in productive activities, some more tangible (for example seeds, agrochemicals, credit) others more intangible (for example husbandry, management, forms of organization, markets). Extension can be applied to any one or a combination of these factors. The main domains in which extension functions are performed include activities relating directly to agricultural production (which have historically received the greatest attention in developing countries); relating to the wider production context (such as the use of trees or water resources in agriculture, wider environmental issues such as soil conservation, health and safety issues, and processing or marketing issues), and relating to aspects of livelihoods beyond the purely economic factors, such as issues of resource access, risk, and voice.

The more tangible the factor and the more directly related to production, the more likely is extension to be concerned with training and advice on technologies (what is often referred to as 'transfer of technology'). On the other hand, the more intangible or related to the wider context, the more is extension about 'facilitating' ways of doing or organizing. In reality, there are various positions between these two extremes, but it is worth noting that, whilst 'transfer' approaches may work well for production activities, 'facilitating' approaches are likely to be more appropriate in efforts to improve wider aspects of livelihoods.

Globalization is impacting the transfer—facilitation continuum in new ways; as producers find it necessary to adopt both new practices and new institutions if they are to take advantage of new markets. An example of this is the question of how extension should relate to regulatory and quality control institutions. It has long been recognized that extension agents should not become directly involved in control functions. Policing and facilitating have been assumed to be mutually exclusive tasks. Nonetheless, the importance of meeting quality standards and obtaining certification for export markets is growing. The poor need institutional as well as technical support in order to gain access to markets for organic and dairy products for instance, and to sell their products through supermarkets. Examples are emerging where the same 'extension agents' are ensuring that farmers meet these standards and also advising them on how to do so. The studies in this volume show that traditional extension services are rarely up to meeting these challenges, but that private and non-governmental organization (NGO) actors are increasingly involved, often in ways that blur former distinctions between regulatory and advisory functions.

An analysis of the scope of extension does not automatically generate formulae on how to design and manage extension services. It is now widely accepted that no single actor or agency is best placed to offer the range of services that poor farmers require. There is a need for a plurality of extension actors to support producers and the rural poor by undertaking various extension activities. Many of these agencies may not consider themselves as 'extension agencies' *per se*. These 'new' extension agencies may be individual traders, producer organizations, micro-finance institutions or enterprises contracting farmers to supply to supermarket chains. Our studies have shown that they are already incorporating extension-like functions in their work. A gap remains, however, in ensuring that the 'old' extension services reorient their own scope to support and make space for the newer actors.

The organizations best placed to offer extension services can largely be guided by the roles defined for public and private actors and agencies. Table 1.1 sets out some of the principal models for structuring and organizing extension, ranging from 'pure' public sector models to private, with collaborative and contracting mixes in between. These 'in-between' modalities are gaining greater attention along with a growing realization that public sector funds can be managed in different ways. For instance, they may be used to contract private sector actors (Rivera and Zijp 2002). A more radical approach is to put funds in

TABLE 1.1

Structure and Organization of Extension

| | 'Pure' public sector model | Collaborative model | Contracting model | Private sector model |
|---|---|---|---|---|
| Planning, management, and funding | By public sector | Government and other agencies (producer organizations; service-providing NGOs) jointly responsible for agreeing on priorities, modes of operation etc. Funding generally from government | Generally by government (though (b) may be evidence of a breakdown in the capacity of the government to plan and manage) | Include:<br>• Commodity-based<br>• Area-based (example for an irrigation scheme)<br>• Farmer-to-farmer<br>Private funding is a dominant characteristic |
| Implementation | By a public service | Agreed division of responsibilities for implementation | (a) government contracts in private organizations and individuals to implement extension to agreed specifications;<br>(b) government staff contracted in or part-funded by private agencies | Generally by private (for-profit or non-profit) organizations |

*Source:* Authors.

9

the hands of producers themselves with the intention of enhancing the accountability and client-orientation of extension agencies to users. Various voucher schemes provide the few fully-analysed examples (Swiss Centre for Agricultural Extension and Rural Development 2000). Important questions exist about the extent to which contracting and vouchers have been effective to date, what preconditions need to be in place for schemes of this kind to function more effectively in the future, and how realistic it is to expect such conditions to be put into place.

The studies in this volume cover the full range of structural and organizational possibilities in Table 1.1, but do not provide any 'blueprint' answers. Experience tells us that there are many pitfalls to simplistic notions of the transferability of institutional arrangements. All this points to the need to consider the eventual viability and desirability of any 'institutional fix' based on a profound analysis of the field level incentives and disincentives for institutional reform. This is especially true in the search for pro-poor extension, as our studies have shown that many current reform efforts have served to further marginalize the poor.

## CRISIS IN PUBLIC SECTOR EXTENSION SERVICES

Agriculture comprises the overwhelming majority of rural private economic enterprise. What, therefore, is the role of the state in the context of essentially a private industry? Those taking a neo-liberal view generally argue that the state should only intervene in order to correct market failures (see Box 1.2), or to compensate for them where failure is chronic, for instance, by providing 'public' goods and services. The studies in this volume explore the roles of different actors in extension by first asking how far privatization has fulfilled its promise of mobilizing private resources. Where gaps exist, questions are then raised about how the private sector may be augmented by government and donor funding in a realistic manner, focusing on public goods.

In addition to addressing market failure, governments can also legitimately engage in pro-poor distributional activities, the aims of which may not be to directly promote growth. Most rural development policies are based on the assumption that market-driven growth will lead to a reduction in poverty. This assumption may be debated, but it is probably valid in countries with low levels of inequality

Box 1.2

Features of Market Failure

The characteristics of market failure include:

- public goods, from which private suppliers cannot appropriate benefit streams. In a pure market situation, goods or services of this type would, therefore, be under-supplied. Public goods typically include training and advice on subsistence crops, and on soil and water conservation. What constitutes a public good in some settings may not be one in others. For instance, the products of research geared towards small farmers may be patentable in some countries but not in others;

- externalities, which have positive or negative impacts not encompassed in the monetary cost of the activity. Of major concern are external impacts on the environment—attributable to poor soil and water conservation, deforestation, and so on. To 'internalize' the full cost of these effects would reduce the activities concerned to socially desirable levels;

- indivisibilities and increasing returns, which create barriers to the entry of new companies, and contribute to a monopoly supply of services. These are typically found in the provision of large-scale infrastructure such as irrigation reservoirs and are, as such, less relevant for extension;

- high levels of risk associated with particular activities, which are not adequately covered by insurance provision. Typically, high risk may discourage smallholders from producing higher-value crops or from keeping large units of livestock, such as cows;

- inadequate information on which to base rational economic decisions.

*Source*: Carney and Farrington (1999).

where other policies are in place to ensure that growth is inclusive. Pro-poor distributional policies include: (i) the provision of safety nets for those temporarily or permanently unable to engage adequately in economic activity, and (ii) strengthening of the 'voice' of the poor in determining their own future, for instance, through the creation of local-level resource user-groups. Such policies may also include measures to promote access by the poor to a range of assets, markets, and services, within and beyond agriculture. There is a growing view that adopting a purely neo-liberal stance undervalues these functions. Policies that focus solely on growth may be driven more by political motives of 'busting the state' (Girishankar 2000) than by analysis of where public expenditure is most justified. The studies in this volume show that, whereas a few years ago agricultural policies generally disregarded the need for complementary policies, poverty reduction strategy papers (PRSPs) and other new instruments have provided a basis for a broader view. That said, there is still limited evidence that these conceptual frameworks are having a concrete impact on specific policies related to agricultural production; there appears to be a lag between the new policy commitments that are being made to poverty alleviation and their impact on specific extension strategies.

After independence, most poorer countries followed 'statist' models of technical change in agriculture and natural resource management, with the public sector dominating or monopolizing the supply of inputs, credit, research, extension, and marketing systems. This led to the creation of large public administrations, hierarchically structured and staffed with permanent employees fulfilling various roles and functions that were thought to maintain a broad impact and equity in the treatment of clients. Financial crises and a common feeling that these public administrations were too inflexible and unresponsive—with a high cost bringing insufficient benefit—led to a fundamental rethinking of the role of the government, along with that of other actors and agencies, starting in the early 1980s.

Regardless of how one views the role of public expenditure, it is clear that if publicly funded extension activities are to be sustained, it is imperative that priorities are made explicit, are justified, and are properly implemented. Public extension services in many countries have simply 'gone bust'. This is often the result of their being repositories for an assortment of political objectives and programmes. When new ideas appeared, it was simply assumed that 'they can do it', without a policy framework to suggest whether they 'could' or 'should' do it. In sub-Saharan Africa, vacancies have been left unfilled and operating budgets approach zero as donors' interest has swung away from extension in particular and agriculture in general. In Latin America, large parts of the formerly public extension services have been privatized or have simply disappeared. In South Asia, extension services are generally less donor-dependent, but face chronic problems of weak management, unfilled vacancies, and formulaic approaches. Business as usual is no longer an option, even if strategic thinking is still frequently lacking.

## PRIVATE SECTOR ROLES

Much of the discourse for reform has been underpinned by welfare economics that helps define whether the 'nature of the good' is more public or more private. The neo-liberal project suggests that to facilitate the private sector in handling private goods and services, and to limit the role of the public sector in correcting market failure and providing public goods and services to help achieve new goals of greater efficiency, improved effectiveness and enhanced accountability are required. Such analysis has been used to demonstrate that many of the functions related to technical change in agriculture being undertaken by the state were essentially private in nature and were, therefore, better offered by the private sector. The drive towards privatization has become an almost inevitable development strategy, both by transferring some services into private ownership and by internalizing private sector principles in public administration, that is 'new public management' (NPM) (see Minogue *et al.* 1998).

It is principally in the areas of service provision that international finance institutions, and, in response, governments, have been promoting privatization. In its full sense, privatization involves the sale of former state enterprises (example, in input supply, marketing, and processing) to private sector organizations. However, also widespread is the introduction of private sector practices within publicly-run operations, such as charging for services in order to achieve a degree of cost-recovery. Efficient market performance is premised on adequate competition among enterprises, adequate access to information, and known risks.

Clearly, many of these conditions do not hold in much of the developing world. The studies in this volume on the impact of such reform programmes on extension indicate that, whilst some conditions might hold in favoured areas close to urban centres, the basic assumptions about the role of the private sector do not hold elsewhere. The result of the reforms has been that the majority of the rural population is subject to high degrees of private monopoly, or (once the state has withdrawn) no services at all. There are no signs that an array of private service supply firms is lining up to compete over the paltry purchases of a poor pastoralist in war-torn northern Uganda or in isolated villages high in the Andes. The supply of improved inputs (high yielding seed varieties, agrochemicals, and machinery) to remote areas has been hit particularly hard, as has the provision of marketing and processing services, and of finance for the purchase of inputs as well as the marketing of produce.

Within reformist agendas, two linked roles are defined for the state: one is to take measures to correct market failures, the other is to provide (or commission the provision of) goods and services where market failure is persistent. As emphasized earlier, governments can also legitimately engage in pro-poor distributional activities as an attribute of social policy, that is, without necessarily having the immediate aim of promoting growth. While it is largely accepted that the state can have a role in providing extension where there is market failure or distributional objectives, the dominant discourse suggests that this does not necessarily mean that it needs to deliver services (see, for example, the Common Framework of the Neuchâtel Group—Box 1.1). Public finance for private delivery approaches is being promoted as an alternative model which seeks to solve the problem of poor performance of public services while ensuring that these services do not disappear altogether (Kidd *et al.* 2000).

'Contracting out' is one institutional mechanism emerging from the NPM strand of reform—being presented as the key channel for public financing of extension in several African countries (example, Uganda, Ghana, and Tanzania)—following the demise of training and visit (T&V) systems (see Table 1.1). Public extension services are being replaced by private service agencies, bidding for contracts from the public purse (Rivera and Zijp 2001). Such initiatives follow in the wake of experiences in Latin America, even though the evidence there is far from clear or convincing (Bebbington and Sotomayor 1998). In the studies in this volume, Nicaragua is the clearest example of a concerted effort to move in this direction. The quality of results of these new structures is a matter of debate, but it is quite clear that these models have had limited effectiveness in reaching the poor. The cost per head of serving the poor is far higher than that of reaching the rich. The new extension providers, that are assumed to be more efficient at cost cutting than the state, have not been effective in concerted targeting of poor and isolated farmers. Such clients are expensive to serve, and little attention has been given to the creation of incentives that might encourage private firms to search them out.

Indeed, reality is often more complex than the NPM discourse acknowledges. In Uganda and other countries, the reverse of the public finance/private delivery structure has emerged. Without operational funds, the public service has under-utilized technical expertise.

Extension agents—sitting in offices without the resources to do any practical work—have made themselves available to private sector actors (either for-profit organizations or NGOs) with access to operational funds but lacking in technical expertise. This leads to a private finance/'public' delivery approach that few had predicted.

While our research has revealed few easily replicable models, it has shown that the design of public service provision in general (and extension in particular) demands a more empirical and iterative approach to basing efforts in local institutional arrangements. There are a growing number of examples of public–private interaction in the provision of services that go beyond more simplistic notions of the role of different agencies. Potentially, these new configurations are particularly important in strengthening the 'facilitation' role of extension. The use of the public–private dichotomy has been criticized for being too crude and simple (Ostrom et al. 1993). Sustainable institutional reform demands analysis of the incentives that influence actors at all levels (Ostrom et al. 2002). The institutional landscape is becoming much more complex under the driving forces of globalization and democratization. Indeed, some of the more interesting examples of service provision point to a 'blurring' of the public and the private (Tendler 1993).

## PRODUCER ORGANIZATIONS

Producer organizations have achieved prominence in the discourse on extension reform and are expected to have an increasingly important role in the provision of services to their members (see Carney 1996). Many producer organizations seem to provide a balance between the growth potential of a market-orientation and some distributional objectives associated with collective action. They are sometimes assumed to be a panacea for the problems caused by both inappropriate governmental involvement and market failure. But concerns are also being raised about their social inclusiveness and reach. While producer organizations will undoubtedly be an increasingly important component of the institutional landscape in the future, it cannot be assumed that poorer producers will automatically gain equal access to resources or benefits through such organizations.

Furthermore, group formation is often used as a part of an extension approach, thereby increasing coverage and reducing transaction costs for addressing common problems faced by producers. There are, however, conflicting trends. There is some evidence to suggest that this strategy runs contrary, even blindly against the tide of globalizing forces which seem to be contributing to an erosion of social cohesion in many societies, through migration and the monetization of social relations. Historical factors greatly influence how farmers perceive the costs and benefits, and the opportunities and risks of collective action. Past history with government dominated co-operatives, for example, in Nicaragua and Vietnam, have left a distrust towards the promotion of co-operatives, even though there is an increasing awareness of their importance.

Trends towards the privatization of research and extension in the North have drawn even more attention to the role of producer organizations, and many have begun to assess the prospects of a greater role for them in bringing about technical change in developing countries. Strong producer organizations have indeed sprung up in a number of developing countries, but these tend to be rooted in very specific conditions. In some countries, the commodity base is the most important (for example cotton in Mali, groundnuts in Senegal). In Nicaragua, new types of co-operatives are being formed to manage the collection, sorting, and transport of perishable vegetables in order to sell to the supermarkets that are increasingly dominating the market for agricultural produce. In Vietnam, co-operatives have re-emerged after a period of decline to provide services that the private sector is ill-equipped to handle. In both Uganda and Nicaragua, the demands of certification of organic products are resulting in new forms of co-operation.

In all these examples, extension activities are included in some ways, but they are not a dominating factor in why and how producers work together. Despite positive examples, some research has shown that rhetoric often runs ahead of reality with regard to expectations that producer organizations will provide a major vehicle for promoting technological change. A more sober assessment of the prospects for producer organizations to support technology change for and with small-scale, mixed (and semi-subsistence) farmers (Carney 1995) argues the following:

1. producer organizations play very limited roles in supporting technical change; they are constrained by combinations of: statutory barriers (which may rule out this kind of activity); lack of resources; the complexity of members' needs (especially where small scale, mixed farming is concerned) and poor internal communication; there are also questions over how far they represent the low income farmers;

2. producer organizations are often represented on the governing bodies of research centres, but in reality have little leverage to ensure that their views are heard, even if they have been able to formulate a single set of views to cover the needs of all their members;

3. these factors, when combined with the common view within producer organizations that other areas offer quicker returns for less effort (such as lobbying for more favourable prices or regulatory environments), means that technical change is given low priority by most.

In short, producer organizations may have some role, in specific conditions (for example, market-oriented commodity production), in brokering technical advice for their members. In this sense, they form a part of the plurality of organizations working on extension. However, it is less realistic to expect them to make a substantive contribution to the reduction of poverty or vulnerability: small-scale mixed, semi-subsistence farmers are a highly diverse constituency, potentially making a multiplicity of demands and unlikely to have the funds to even pay for membership. Therefore, they hold little appeal as potential members. Even where they are members of producer organizations, such organizations are unlikely to command the resources to meet more than a fraction of their varied demands.

In the light of these conclusions, the role of producer organizations *vis-à-vis* extension and poverty must be considered in relation to questions of the impacts on poverty of supporting relatively well-off farmers. By helping to expand large-scale farming and to enter export markets, the kinds of technological change that producer organizations promote may often primarily impact on the poor by job creation. Producer organizations are significant actors in agricultural change more generally, and thus cannot be ignored in efforts to strengthen extension for the poor, but should not be assumed to be ready or effective vehicles for either managing extension or for helping the poor to draw upon services.

## LIKELY IMPACT OF GLOBALIZATION

In general terms, globalization refers to the growing interdependence of the economies of the world, and comprises the following:

1. substantial increases in capital movements—with average daily foreign exchange transactions having increased from $15 billion in

1973 to $1.2 trillion in 1995, and international capital movements now exceeding trade flows by a factor of sixty (Sutherland 1998);

2. rapid growth in world trade, with a likely doubling in trade every 12 years. Low elasticities of demand for primary goods mean that the share of agriculture in total world trade has been falling; but, even here, agricultural exports have grown at twice the pace of world agricultural output (WTO 1999);

3. internationalization of production through the growth of multinational corporations; the prospects of more rapid spread of new technologies such as genetically modified organisms providing both threats and opportunities to the rural poor;

4. the declining relative costs of international transport; and

5. the rapid spread and declining costs of telecommunications and information flow associated with the information technology revolution.

Despite these seemingly overwhelming trends, not all market factors have been liberalized. In the case of labour, for instance, largely because of immigration policies, only some 2.3 per cent of the world's population lives outside its country of birth (World Bank 1999). Further, it is only private capital movements that have been internationalized. Public capital in the form of development assistance declined by 0.7 per cent per annum in 1988–97 in real terms (UN 1999), with aid to agriculture falling by almost 50 per cent in real terms during the period 1986–96 (Pinstrup-Andersen and Cohen 1998).

Finally, there has been limited progress in specifically agricultural liberalization. The Organization for Economic Co-operation and Development (OECD) countries, for instance, protect their own agriculture to the tune of US $1 billion per day, and pressures in both the USA and European Union (EU) during 2002 were towards continuity and, in the US case, even higher protectionism.

Trade theory predicts that, since developing countries have a relative abundance of unskilled labour, freer trade should increase the demand for exports embodying large inputs of unskilled labour, thus increasing employment, raising wages, and reducing poverty. This would, at first, appear to indicate that, although there may be declining potential for small-scale producers to retain access to local markets, increasing wage labour opportunities should provide alternative livelihoods. This is not necessarily the case. In a comprehensive effort to assess the current and likely impact of globalization on the rural poor, especially on those relying principally on agriculture, Killick (2000) concludes that increased world demand appears to be most buoyant for products embodying skilled labour and relatively advanced technology. Globalization is likely to have little positive impact, and possibly a number of negative ones. He attributes this to five sets of factors.

1. extensive continued protectionism in OECD countries, which has severely limited the amount of agricultural trade liberalization that has occurred;

2. limited liberalization within developing countries, with the persistence of anti-agricultural biases (Schiff and Valdes 1998);

3. biases in technical progress in favour of capital and skill intensity, and towards commodities which do not easily lend themselves to production by small farmers in remote areas;

4. a wide range of factors (many associated with market failure) preventing the rural poor from responding as well as they might to emerging market opportunities—including scarce market information, poor infrastructure, weak institutions, and inadequate access to assets such as education, land, water, and finance; and

5. the high proportion of rural population, especially in Africa, that finds it difficult to be economically active, including the handicapped, the aged, orphans, refugees, and female and child headed households. In many settings, this is accompanied by the weakness of modern safety nets, and the erosion of traditional ones.

Killick concludes that in many developing countries the welfare of those involved in agriculture will be increasingly dependent on the sale of their labour, and so determined by the efficiency of labour markets and increasingly linked with the wider development of the economy and the growth of non-farm employment opportunities in rural areas.

Kydd et al. (2000) broadly concur with Killick's analysis, but add two further areas of difficulty of particular relevance for extension.

1. New GM technology is spreading more rapidly because of globalization, but has attributes that make it easier for those with adequate skills and access to markets to acquire it. The capacity of GM technology to quickly flood markets means that the disadvantages of late adoption are accentuated, and so there will be increased returns

to any effort by extension which narrows the gap between early and late adopters.

2. Access to knowledge, credit, and markets is becoming more crucial as globalization gathers pace. This gives particular advantage to (usually) larger farms located in favoured areas, making outdated some of the conventional wisdom concerning the superior efficiency of small farms and, again, raising the returns that can be achieved when effective extension and other services are provided to small farms.

Our studies have shown that these dangers are proving real, particularly in relatively accessible areas that are most impacted by globalization. Furthermore, there is little indication that extension and agricultural development planners are taking into account these new trajectories when deciding how to support the livelihoods of the rural poor. Rural labour markets are usually considered 'somebody else's business'. In fact, efficiency and competitiveness are generally automatically equated with a shift from labour-intensive to capital-intensive ('modern') technologies. Poor farmers are implicitly re-designated as 'excess populations' that are better-off finding alternative livelihoods.

The primary exception is in the isolated or conflict-prone areas where even 'modern' technology is obviously on the retreat. Use of fertilizers and high yielding varieties is declining in places where the 'market' fears to tread. The notable exception is illicit production of narcotics and smuggled goods, where global entrepreneurs have found a certain 'comparative advantage' for agricultural production in areas outside the reach of the government. Apart from narcotics though, there are few examples of innovative approaches to addressing this downside of globalization. Despite rapidly increasing expenditures on agricultural rehabilitation to address the collapse of agricultural production in the face of drought and conflict, these questions are rarely on the extension agenda.

## VULNERABILITY

Pro-poor extension directed at the poor can (and should) be justified on more than its contribution to increased production. In addition to providing poor people with the information and institutional support that they need to increase their wealth and well-being, and thereby 'escape' from material poverty, extension can also help them to mitigate risks to their livelihoods and thereby provide a higher degree of stability. This involves reducing vulnerability, rather than accumulating assets.

Asset accumulation has traditionally dominated most development efforts directed at poverty alleviation, and extension support is no exception. 'Risk aversion', implying the desire to reduce vulnerability rather than accumulate wealth, has traditionally been perceived as a central 'problem' for those involved in extension. Farmers have feared the potential negative consequences of following the advice they receive from extension agents. When the perspectives of poor people themselves are analysed, it appears that risk reduction is frequently given priority over asset accumulation. They are ready to innovate, but they want to do this by learning how they can become better at being 'risk averse'. There are many examples where poor people have adapted initiatives that were originally intended as means to escape from material deprivation, to instead protect the assets that they have. For example, micro-finance programmes that have been established with the aim of asset accumulation have increasingly been found to be used by the poor as a source of capital to deal with livelihood shocks, that is, as *de facto* insurance policies (Matin *et al.* 1999).

Fears of destitution and acute suffering may often be more immediate concerns than getting rich. The changing world economy is contributing to a change in the nature of these fears. Livelihood shocks are becoming more common because of natural disasters, conflict, and market turbulence. Economic problems in urban areas have an increasing impact on rural people, who have become more dependent on markets, part-time employment, and remittances from relatives in towns. Greater interdependence has increased the transmission of economic shocks and conflicts across borders. Capital can move to avoid such shocks, but the poor are far less mobile and, therefore, far more vulnerable. Coping strategies have been weakened along with declines in the subsistence economy. Commercialization has meant that help from wealthy neighbours is less forthcoming and horizontal ties are subject to covariate risk, implying that when need is greatest it is unlikely that other poor people will have the capacity to help (Devereux 2000).

Whereas in the past, a low-risk strategy was often synonymous with a focus on subsistence staples (and thus a problem for extension services geared towards cash crop promotion), poor people are increasingly struggling to reduce their vulnerability by diversifying their livelihoods within and beyond agriculture (Berdegué *et al.* 2000; Bryceson 2000). By keeping their eggs in several baskets, poor people are able to take advantage of emerging opportunities and minimize losses from weak markets. When most poor people no longer have the opportunity to meet their subsistence needs through their own crop production, diversification is increasingly attractive as a risk mitigation strategy.

Further, the line between asset accumulation and vulnerability reduction strategies is constantly shifting and is closely related to the availability of services. If poor farmers lack confidence in access to information, credit, and marketing services, they will naturally adopt a more risk-averse production strategy. Decentralization and the existence of democratic and responsive governmental institutions at local levels may create a sense of greater security, which will in turn encourage more long-term investments. On the other hand, if decentralization and outsourcing of public services results in chaotic and uncertain relations between service agencies and the poor, the reverse may occur. Nicaragua demonstrates both these tendencies. In better endowed municipalities, where privatization has been managed effectively and where governmental institutions function relatively well, there are signs of rapid agricultural innovation and growth. In isolated areas, the opposite is true. Large tracts of land have been effectively abandoned. Investment in beef cattle, for example, has fallen due to uncertain international markets and local insecurity.

For many households in such areas, accessing semi-skilled labour opportunities (through education) on larger farms may be a lower risk strategy than investing in their own plot of land. Migration and access to food for work and other public works schemes have become more 'sustainable' than their traditional livelihoods. Reliance on strategies that were once seen as symptoms of poverty, are now being recognized (at least by the poor themselves) as part of the solution. Reducing vulnerability means making the best of the options available.

The changing nature of links between urban and rural areas is affecting vulnerabilities in different respects (Satterthwaite 2000), that are relevant to extension strategies and priorities:

1. vulnerability is being reduced for those near urban areas due to increased possibilities for diversification, commercial horticulture, and livestock production as well as alternative employment;

2. vulnerability to traditionally urban hazards (pollution, crime, etc.) is seeping into rural areas, as is vulnerability to turbulence in the urban economy; and

3. the traditional vulnerability of isolated rural areas is increasing as they are effectively excluded from new economic structures and as many states reduce or withdraw the provision of public services in expensive and non-dynamic hinterlands.

In many areas of the world, the picture is even bleaker. Development is no longer on the agenda. A combination of ecological, agricultural, and social systemic collapse has meant that disasters are no longer temporary disturbances in the grand march of development, but are instead indicators that 'durable disorder' is taking hold. Some of these are in arid hinterlands (example, the Horn of Africa and Afghanistan), where conflict has been endemic for centuries. Other places were until recently productive zones, but inappropriate agricultural systems (Tajikistan) or armed conflict (Central Africa) have pushed back government authority, allowing 'uncivil society' to fill the vacuum. The poor inhabitants of these places have reacted in various ways. Some have adapted cropping patterns to minimize risk (as in Sierra Leone; Richards 1996), many others have joined the drug economy (as in Afghanistan; Goodhand 2000). Smuggling networks, such as in the Balkans, often define the parameters for the new rural economies. As mentioned earlier, globalization's capacity to link local production to global markets has created lucrative opportunities for investments in areas outside formal or legitimate state control (Duffield 1998).

All this creates enormous challenges with uncertain outcomes for those hoping to contribute to reconstruction. The implicit assumption that support to agricultural rehabilitation should be about helping people simply to return to the livelihoods they pursued before the crisis has given way to an increasingly acknowledged uncertainty about what it is that should be rebuilt. As conflicts drag on for decades in many parts of Africa, and as entire production systems collapse, as in parts of the former Soviet Union, the rural poor are no longer farmers simply waiting in refugee camps to return to their farms. Many refugees and internally displaced persons returning to the rural areas of Afghanistan and Angola have little direct experience of smallholder production, having adopted livelihoods as soldiers, refugees, labourers, and slum dwellers. Re-establishing rural livelihoods is not just a matter of going home and pumping in fresh investment capital. Knowledge is key. The need for extension when people return to rural areas is enormous but the capacity to mobilize institutions, whether public or private, to meet these challenges is usually extremely limited.

Even in 'post'-conflict situations, the homes that people are returning to are often scarcely reached by government services. Government capacity to provide minimal basic services is directed to higher priority areas, and civil servants refuse to work in the insecure and forgotten hinterlands. A brain drain draws the commercially ambitious and well educated to more dynamic areas. As already mentioned, one exception is the magnetic attraction of hinterlands to those who wish to take advantage of the absence of formal authority. The peace that has finally come to Afghanistan is bringing with it a massive increase in poppy cultivation. This is more than a return to age-old traditions. The corrupt and criminal forces that currently tend to fill such vacuums in authority are unlike the warlords of old. They are often transnationally connected and have learned to manipulate both humanitarian and development initiatives. This creates special dangers in promoting extension, as the desire to rush in and strengthen the capacity of local institutions may be at odds with the need to be extremely careful in choosing which institutions to strengthen.

## HIV/AIDS

One feature of systemic crisis that is of critical importance for development in general is HIV/AIDS. It is not just a health issue. The agricultural sector is particularly affected (Piot *et al.* 2001). In some of the more heavily affected countries there have been dramatic shifts in the demography of the population. It exacerbates existing problems such as labour bottlenecks, problems of rural women especially female-headed farm households arising from gender division of labour and land rights (du Guerny 1999). In some areas, the combined impact has reached crisis proportions. Agricultural development and food security are key areas that need to be monitored with reference to the impact of HIV/AIDS together with the response of rural populations to the pandemic at national and international levels.

The HIV/AIDS epidemic is of concern for rural policy, though some countries have higher rates of infection in urban areas. Zambia has a prevalence rate of 33 per cent among women in urban areas compared to 13.2 per cent in rural areas. The rates are much closer in other countries (such as South Africa) and a key factor is the amount of movement and interchange between urban and rural areas. One hypothesis is that successful rural development facilitates the spread of HIV/AIDS through the creation of opportunities for migration and employment. It is imperative that extension priorities incorporate safeguards to minimize this. Learning about how to avoid and to deal with HIV/AIDS is a part of learning how to maintain a viable livelihood for many rural households.

The impact of HIV/AIDS on economic growth is such that per capita income growth has been reduced by 0.5–0.75 per cent per year in countries with adult prevalence rates of 10–13 per cent (World Bank 1997). The reliance on labour intensive agricultural production in many poorer economies makes this pandemic particularly important. In West Africa, there is evidence to suggest that HIV/AIDS has reduced cultivation of cash crops or food products, including market gardening in Burkina Faso and cotton, coffee, and cocoa plantations in Côte d'Ivoire. A study in a Tanzanian district suggests that the time spent on farming has shifted radically because of AIDS. Women with sick husbands spend 60 per cent less time on agricultural activities than is the norm. In rural areas, the impact is greatest when the primary income earner dies, or when relatives who return to the village are no longer able to send remittances from urban employment, and, in fact, absorb resources by needing to be looked after in the village.

Although information is scant at present, it seems that the impact varies in different farming systems given the variation in labour requirements of crop and livestock enterprises and seasons (USAID 1996; UNAIDS 2000). Small-scale crop production relies on family labour and a number of other household activities such as food processing and home maintenance can be significantly affected by the loss of family members. Hiring labour has been a common response. But survival strategies of the poorest households, such as the sale of assets and use of any savings, make them particularly vulnerable following the death of a household member. One response has been to shift towards livelihood systems that incorporate less labour-demanding crops and enterprises. This disease can, therefore, have the most significant consequences on the poorest households who are least able to cope with illness and death. Children are amongst the worst affected as many are orphaned or have to leave school to look after sick relatives. The changing demographics, with a greater proportion of older and younger, is clearly a challenge to extension agencies that must consequently reconsider their priorities and approaches.

Agencies offering agricultural extension are also feeling the impact of HIV/AIDS among staff. The effects are felt in reduced productivity among those infected and those taking time off to attend the funerals of colleagues and family members. There are also losses in staff morale and increased costs in terms of employee benefits and replacement. Further, there has often been high investment in human resource development within extension organizations, on staff education and training. Indeed, it has been suggested on occasions that the risk of investing in human capacity building has increased significantly with the HIV/AIDS pandemic and such strategies must be questioned accordingly.

## DECENTRALIZATION

Trends towards decentralization are having a profound impact on privatization and changing organizational roles in rural policy and practice. Through the 1980s and into the 1990s, centrally led public sector institutions were increasingly seen as being out of touch with the needs of farmers. Decentralization has, in many quarters, almost come to be seen as a panacea in terms of increasing proximity, relevance, autonomy, participation, accountability, and democracy. Expectations have been raised that extension services could be rejuvenated in the process, and there are some examples (as in Vietnam) where this has happened. On the whole, however, the impact of these changes on extension has been mixed at best.

In considering decentralization, there is a danger in making sweeping generalizations. The story is complex and often underpinned by strong political discourse in favour of one 'solution' or another. Ostrom *et al.* (1993) note that there are often more decentralization claims than facts. As with privatization, this is an arena where care must be taken to look beyond the prevailing rhetoric. In aiming to understand better the dynamics of decentralization, it is commonly acknowledged that the concept cannot fit a centralization–decentralization dichotomy or even be seen as a continuum between two poles. Rather, it needs to be located in and analysed according to multiple dimensions (see, for example, Conyers 1985; Ostrom *et al.* 1993). Conyers (1985) highlights five dimensions characteristic of all decentralization efforts. They are: (i) the functions over which authority is transferred; (ii) the type of authority or power transferred with respect to each function; (iii) the level(s) or area(s) to which authority is transferred; (iv) the individual or agency to which authority is transferred at each level; and (v) the legal or administrative means by which authority is transferred.

In terms of agricultural development, what is clear is that local government is unlikely to play a strong role unless it is given the mandate and resources to carry out (or influence) particular functions. In practice, decentralization strategies are commonly described in terms of political, fiscal, and administrative components. However, it cannot be assumed that these components will be in harmony and there are cases in the agriculture sector where political decentralization may not be met by the necessary fiscal decentralization. Populist calls for letting the farmers decide are not always accompanied by the cash to allow them to act on their decisions.

This does not seem to be a problem in Uganda, for example, where district councils have the responsibility for both extension and budgeting of extension. Pro-poor agricultural development has been a cornerstone of much of the process of planning new decentralized structures. Difficulties have arisen due to local government enthusiastically raising its own income through vastly increased fees and taxation on agricultural products (Ellis *et al.* 2001). Farmers (especially the poor) are being pushed back into subsistence as their would-be profits from commercialization have disappeared into the pockets of newly empowered local administrators.

In Colombia, the management of agricultural extension by the 'municipalidades' has yielded positive results. Local governments are expected to use their resources for promoting technological change in agriculture. In Bolivia, decentralization has given local government the freedom to redirect resources away from extension related activities, as other activities are given fiscal priority. The situation is even worse in the poorer areas of Nicaragua, where local governments have few resources to work outside the municipal towns, and no experience in rural development issues. They continue to focus on urban services since they lack the capacity to shoulder their (officially) expanded mandate beyond the edge of the towns. An exception to this are the larger municipalities that experienced widespread destruction due to landslides and flooding after Hurricane Mitch. In these areas, a realization that even the urban population will suffer if natural resource management is neglected has inspired a new interest in these issues.

In India and Vietnam, agricultural development is definitely on the agenda of local government. In India, local government has access to an array of different rural development and poverty alleviation programmes and funds. For example, district levels and below have been mandated to organize and fund local-level planning and related service delivery in a total of 17 distinct natural resource spheres, including microwatershed rehabilitation which attracts central government expenditure of some US$ 500 million per annum. The implementation of these programmes has been problematic due to widespread politicization and corruption (Saxena 2001). While providing a significant degree of resources for extension, these efforts have also drawn extension into a complex (and not necessarily pro-poor) web of local power relations.

In Vietnam, the realization that agriculture is the foundation of local economic development has created strong interest in supporting extension. Decentralization has also enabled and encouraged different local actors—government, mass organizations, and co-operatives—to find creative and pragmatic approaches to collaboration on extension activities. This co-operation draws on the strengths and interests of local actors, and often focuses on locally identified priorities. Special, centrally funded poverty alleviation funds and, in some areas, the need to mitigate recurrent flooding, have served to further galvanize these efforts.

A dynamic civil society is commonly regarded as an indispensable component of strategies for democratization and decentralization. But in many countries, and particularly in the poorest regions, these organizations remain weak; formal producer organizations are few in number and rarely represent low-income farmers. With some notable exceptions (for example, in Bangladesh), service-providing NGOs are few and far between, and where they do exist, their high dependence on foreign funds may call their local accountability into question. Also, traditional institutions have long proven able to manage common pool resources such as rangeland, forest and water effectively, but in many areas rapidly increasing population pressure on these resources has caused their decline; and only rarely have they been effectively incorporated within the changing patterns of civil society or replaced by new organizations (example Ostrom et al. 1993).

Despite the failure of decentralization to live up to its grand rhetorical objectives, it is nonetheless at the crux of the political process that will inevitably determine whether agricultural services can be restructured to enable farmers to draw upon what they need. It impacts on whether people have a strong voice in decisions on development that affect them. It affects whether minorities can assert their rights. It determines if and how systems are put into place to demand accountability from the public administration and so reduce the scope for corruption. In all these factors, decentralization can either make extension more pro-poor or less.

While there has been a ground-swell of public demand for local control, the high level of donor promotion of and support for decentralization has compromised 'ownership' of the process in many countries. Line ministries often show symptoms of inertia with regard to decentralization, with those ministries responsible for extension often arguing that they have been the most decentralized (at least in the sense of deconcentration). There seem to be few examples of centralized ministries making 'good exits' within a decentralized framework.

Indeed, Palidano (1999) draws on Tendler (1997) to demonstrate that public action is often a matter of the intertwined actions of centralized and decentralized levels, and questions the predictability of policy assumptions regarding the effects of decentralization on public service provision. Simplistic notions of policy transferability (of which there seem to be many related to agricultural extension) are challenged by real-world lessons in contingency which show that experimentation and eclecticism are key. Certainly, a common challenge is to balance central control and local autonomy. A re-drawing of boundaries between what is best done locally and what requires central leadership is important, as is the provision for cross-learning among decentralized administrations, and between these and central government.

Poverty reduction is becoming an overarching policy objective of government in many poorer countries and this has implications for interactions between central and local government. There are often concerns about 'elite capture' within decentralized structures, and the lack of priority that elected local governments tend to give intangibles such as extension in contrast to, for example, rural roads, which are a more convenient 'objectively verifiable indicator' in the political arena. Even if public support for extension is a justifiable strategy for poverty reduction, it may not always survive local political processes.

## IMPLICATIONS FOR EXTENSION

Pro-poor extension must be constructed around a realization that the livelihoods of the poor are based on their roles as producers, labourers, and consumers, and that technological change in agriculture will inevitably impact on the ways that poor people combine these strategies. Pro-poor extension is thus about promoting more positive outcomes for the poor in all three aspects. This broad interpretation of the scope of extension in relation to poverty is consistent with recent thinking on the nature and purpose of development itself. Amartya Sen (1999) has proposed freedom of choice as a basic criterion of development. Increased productivity, therefore, is only a part of development. A more fundamental aspect is the ability to participate in the market through ready access to information, freedom of movement, and access to the socio-political and institutional infra-structure that underpins the market. When development is described in these terms, extension clearly becomes a central issue. The poor need choice and rural policy can have a pivotal role in facilitating freedom of choice, with extension providing a key support function to help people make better choices and act accordingly.

This in no way implies that access to information is the primary obstacle to poor people's ability to choose their livelihoods. Information is bound together in the political and economic structures that guide, not only immediate production decisions, but also the wider questions of whether the poor have the power to pursue their chosen livelihood strategies. By placing the information role of extension in this context, the clients of extension become more than mere consumers or customers (as in new voucher and contracting schemes), or co-producers (as in farmer-to-farmer approaches), but citizens as well (see Girishankar 2000). The future of public sector support for extension is more dependent than ever on a capacity to strengthen the institutions of the poor, not only in relation to production, but also more widely in relation to access, vulnerability, and voice.

Related to this is the question of how the role of government is best defined with respect to extension. The role of government in development is not only to make good market failures, but also to take on a wider distributional mandate. This is certainly consistent with the involvement of extension in strengthening institutional capacity, but it can apply equally to the provision of information, advice, and training. In the changing world, access to information is coming to be seen as a basic social policy objective (ODI 2000). Extension can, therefore, have both economic and social goals, and so be judged on the principles of both economic and social policy. If not, and if extension continues to be judged on purely economic terms, then triage will become the order of the day. Agricultural services will become increasingly focused on those wealthy enough to engage directly in globalized markets, and poverty alleviation will be left as 'somebody else's problem'. Some states and donors have essentially written off the agricultural livelihoods of many of the poor as being simply non-viable (Bebbington 1999); the implications of this must be more explicitly acknowledged if a coherent and transparent approach to pro-poor extension is to be found. Even in these states, a pro-poor extension agenda can perhaps be found.

In practically all countries, the long-term trend is that of a decline in relative importance of agriculture. Recent evidence suggests that this decline is more rapid than had been thought (Berdegué et al. 2000; Bryceson 2000). Non-agricultural options include accessing wage employment (migratory, casual, and skilled), and engaging in micro-enterprise, both of which relate to the growing importance of rural–urban linkages in the livelihood strategies of the poor (Satterthwaite 2000). The support given by extension for new technology has tended to focus only on the production aspects. However, and not least because of the influence of globalization, greater attention to marketing and processing aspects is now urgently necessary. Extension services must find a new role as part of overall rural development processes by becoming nested within a sound service context and not act as if agricultural production can be seen in isolation.

With regard to employment, questions emerge regarding whether working for a wealthier neighbour may be a more appropriate option than investing in a minuscule plot of land. Economies of scale are changing. While extension still plays an important role in directly supporting smallholders in increasing their efficiency (especially in countries and regions where large-scale production has not taken hold, such as in Vietnam), a two-track approach including support to larger-scale farming may have considerable impact on the livelihoods and well-being of rural households. A closer look at equity and viability issues related to employment generation in medium and large-scale production units is warranted. These questions go to the heart of the basic concepts and assumptions of poverty alleviation. The failure of the trickle-down approaches of the 1950s–70s is well-documented. But much of the promotion of growth in this period took place in a *laissez-faire* environment. Knowledge of the preconditions and instruments necessary for ensuring more equitable distribution of the benefits of growth has greatly increased since then. The studies in this volume show, however, that this knowledge has yet to be applied to rural development (and extension) planning.

Poor people are increasingly reliant on wage labour (Devereux 2000; Berdegué *et al.* 2000; Satterthwaite 2000). Agricultural policy, however, is still often disposed to responding to the demands of larger-scale farmers by subsidizing labour-displacing mechanization. Extension is frequently used (consciously or unconsciously) as a tool in implementing this policy. If it is to be explicitly pro-poor, its role must be assessed in relation to overall policies regarding mechanization and employment generation. The studies in this volume have found few instances where this has happened.

More proactive rural development policies, possibly implemented through extension, can help poor farmers (and especially their daughters and sons) to prepare to meet the increasing demand for skilled labour in agricultural enterprises, and thus find a 'good exit' from smallholder farming. The development of human capital has been cited as one of the most important factors in securing rural livelihoods (Killick 2000; Bryceson 2000). Agricultural education should not be seen in isolation from the broader educational needs of enabling people to pursue a diverse array of agricultural and non-agricultural strategies. The so-called 'yeoman farmer fallacy', the belief that equity can and should be achieved mainly by supporting farmers as producers, has tended to blind development thinking to the diversity of poor people's livelihoods, and to the options for fostering access to wage- and self-employment. Our research has shown that, though the importance of such factors is generally acknowledged, not least in recent PRSPs and other poverty assessments, nonetheless, surprisingly little progress has been made in reconceptualizing extension as an integral part of these new policy frameworks.

## LESSONS FROM CASE STUDIES

In his review of extension in Uganda, Kidd describes how genuine and well considered approaches to poverty alleviation are leading the creation of new rural development policies, and how there is a readiness to apply these new concepts to extension itself. Great challenges remain, however, not least in the conflict torn north of the country and in addressing the myriad of challenges created by the HIV/AIDS pandemic. New opportunities are emerging, partly as a result of Uganda's much hailed adherence to a pro-trade reform agenda, and partly due to its agroecology, which is well suited to diversification. Innovative approaches to extension are being tried by helping input stockists provide better advice to their clients, and through various efforts to use intermediary organizations to help producers better understand and deal with their markets. Information and communication technologies (ICTs) seem to have potential for providing further support to extension personnel in the field, but thus far little has been achieved. Uganda's long-term potential to capitalize on its comparative advantages will, however, be related to whether OECD protectionism remains in force. Further, there are no signs that ways have been found to address 'market failure' in isolated and conflict prone areas of the country.

India has sizeable areas with low agricultural productivity, high incidence of poverty, and with weak integration into markets. Questions are increasingly being asked about the role that public sector extension can play in enhancing the livelihoods of the poor and reducing their vulnerability in these areas. Public sector extension in Indian states started adopting different approaches after the T&V system began to lose popularity. The last decade has seen an increased involvement of private extension providers, but their presence and activities are skewed towards the well-endowed regions. In their chapter, Sulaiman and Holt argue that to perform new roles with wider scope, extension services must change fundamentally, not only in personnel and resources, but also in their basic perceptions and practices as they relate to the role of the state in agricultural and rural development. A newly released Government of India (GoI) consultation document indicates awareness of the broad types of change needed. However, changes in the practice of extension will be slow in India for complex reasons rooted in long-held perceptions about the rural poor, the private sector, and the role of the state. Despite the foresight of the new consultation document, and examples of institutional innovations from within India, the great majority of extension remains publicly funded and publicly delivered. It is geared predominantly to the delivery of messages and (although recently less so) subsidies. Isolated innovations offer insights into potential ways forward for extension in the new millennium, but to reform a system in which there are many entrenched actors across the different states within the federal system is clearly going to be a major challenge. This chapter stresses the fact that reforms favouring the poor are unlikely to be achieved unless agricultural policy towards weakly integrated areas becomes less concerned with productivity enhancement alone, and more with the ways in which increased productivity can be linked to reductions in vulnerability and employment creation. A greater effort in trying different approaches to active partnership between organizations holding complementary skills, evaluating these at the local level and more systematic approaches to organizational learning will be vital to ensure progress.

Bolivia has experimented with different extension models, from the statist institute with research and extension departments through the intermediate users model with a bridging role, to the current proposal of free-market allocation of extension projects to the best bidders. As a result of their past history and poor past performances, currently there is no national extension system in place in Bolivia. Most extension initiatives depend on NGOs, producers associations, and two para-statals: the Foundation for Research and Promotion of Andean Crops and the Tropical Agriculture Research Centre. Decentralization has not provided the expected benefits to extension services and very few municipalities have introduced agricultural technical assistance components to their operational structure. Municipalities' budgets are oriented towards visible (civil) works rather than towards upgrading knowledge systems. The extension component is a major mechanism of rural development strategies missing from the decentralization process launched by the Popular Participation Law (PPL).

Vietnam represents a very different set of developments from the other case studies. It is a strong state with a clear normative commitment to rural development, poverty alleviation, and vulnerability reduction. Extension, rather than being seen as an 'old bureaucracy' in need of reform, is a new phenomenon that has been used by local authorities, mass organizations, and other actors as a vehicle for change. And change has happened. The rural economy has rapidly shifted from subsistence to export production; and diversification away from rice is rapidly expanding. This has brought benefits and challenges for the poor. As a clear target group, they have seen resources channelled via extension to support their development. At the same time, their main 'need' is assumed to be 'modernization', in the sense of more capital-intensive technologies, which may be out of reach for the poorer population. Elite biases are further reinforced by widespread assumptions regarding the ease of diffusion from wealthy farmers to poorer neighbours. Faith in the power of demonstrations and models to elicit widespread adoption encourages extension agents to exclude the poor from their clientele despite a genuine commitment to poverty alleviation.

In Vietnam, poverty alleviation through rural development is in many respects a matter of geographic targeting. Subsidies are provided to help the more isolated and traditionally less dynamic mountainous areas, inhabited by ethnic minorities, to engage in the changing economy. Furthermore, the increasing impact of floods has led to an emphasis, in some areas, on reducing vulnerability through diversification, agroforestry, and a general awareness of the impact of changing land use on risk.

All of the countries reviewed are grappling with globalization, but Nicaragua is the country that has perhaps most thoroughly accepted a need to adapt to realities beyond its borders. As a small, very poor, very indebted country, located very much in the 'backyard' of the United States, Nicaragua has been struggling to come to terms both with market opportunities and the glaring lack of opportunities facing many of the rural poor. Many policies, including the interim PRSP, reflect the emerging awareness and challenges to social protection. Most agricultural service providers in Nicaragua are pessimistic that poor producers will succeed in significantly accessing international markets. A more pressing concern about the impact of globalization is whether poor producers will be able to retain a domestic market in the face of competition from regional imports.

Extension has sometimes been highlighted as an example of Nicaragua's reform efforts, with services increasingly contracted out to the private sector. Consideration of the public goods issues was, for a period, a major feature in the design of new structures. Public goods issues are now receiving less attention, and more attention is being given to generally expanding service provision. The failures of the Government to mobilize a strong response to Hurricane Mitch were rooted in neo-liberal policies that reduced public service capacity.

The institutional landscape in Nicaragua contains a confusing and seemingly paradoxical mix of policies, structures, and priorities. Non-governmental organizations that often trace their roots to leftist initiatives are actively promoting a modest role for the government and stronger market orientation. State bureaucracies, although led by the neo-liberal government, have been slow to adopt a market focus. Also, Nicaragua is a land of projects. Government capacity to use policy as a tool to co-ordinate the mass of projects that together make up the thrust of Nicaraguan rural development initiatives has been limited. 'Projectization' has a profound impact on the nature of institutions offering extension services. Agencies expect to be judged by donors by their potential capacity to undertake different extension tasks, rather than 'correct' service provision slots for state, private sector, and civil society institutions.

## CONCLUSIONS

Discussing globalization, Paul Streeten has noted that 'We are suffering from institutions lagging behind technology' (2000, p. 46). The paradox of public extension services today is that these organizations are specifically entrusted with the task of helping people to overcome the technology lag, and yet they are usually among the weakest performers. Streeten goes on to point out one of the central reasons for 'institutional lag' when he notes that 'creative institutions are not designed on a drawing board, but are the spontaneous responses to challenging situations' (2000, p. 49). The T&V system for extension is a textbook example of the limits of drawing board design to address a dynamic field of work. The studies in this volume look at an array of 'post-T&V extension' and ask whether the institutions involved are starting to catch up.

Reform of the role of the public sector in extension, anchored in a transparent and overriding policy commitment to the reduction of poverty and vulnerability, cannot be undertaken in isolation from the technology–poverty–globalization link. Furthermore, the challenges of reorganization in the context of systemic collapse, uncivil society, and weak governance must not be overlooked, or relegated to simplistic institutional 'fixes'.

Public extension services have a poor record in following changing technologies and markets. It is essential, however, that this failure be addressed as a means for ensuring that the poor have the freedom and opportunity to participate in the global society and economy. The role of agricultural and rural development policy—and of extension as one of its instruments—is to ensure that these means of development are equitably accessible. If new strategies are to be formed, the social and economic implications of different technological trajectories must be brought together. Economic interest and public interest in service provision should be integrated, but not be assumed to be interchangeable (Girishankar 2000).

A new agenda for extension—cognisant of the current trajectories of globalization, yet clearly reflective of the values of supporting the rural livelihood strategies of the poor—requires more than the methodological or organizational fixes that have dominated extension reform efforts in the past. Neither *laissez faire* acceptance of market mechanisms, nor populistic participatory schemes will help poor people deal with the ever-increasing shocks to their ways of life and to access exclusive markets. Difficult choices and prioritizations must be made, based on bringing together an understanding of who are the poor, where are the poor, and what is the role and capacity of governmental, civil society, donor, and private sector organizations. This involves 'getting the policies right'—not only agricultural, but also the wider range of rural development policies that impact on the poor in health, education, microfinance, infrastructure, and so on—and forging links between them.

A theme running through many of these choices is the question of triage. Should one strive to reach the so-called poorest-of-the-poor in isolated areas who are increasingly excluded from the international economy, even if their prospects for either subsistence or market production are meagre? If not, what are the consequences, not only for these people, but also for the broader society? Can appropriate combinations of safety nets and enhanced production opportunities be created? Can existing trends be supported to achieve 'good exits' for the millions who will leave agriculture over the next decade? Is this an important future role for extension—helping people make better choices in the face of structural change? The case studies in the following chapters present key examples of the decisions that need to be addressed in relating extension to poverty and vulnerability. These studies also show that extension is just one (and often minor) component of a range of rather staggering challenges facing reform of rural and agricultural development policy.

## Endnotes

1. There are strong arguments that political capital should be included as a sixth asset (Baumann 2000).

2. Agriculture here is broadly defined to include not only crops and livestock, but also the related management of the wider natural resource base (water, soils, trees...).

## References

Adams, M. (1995), 'Land reform: New seeds on old ground', *Natural Resource Perspectives 6*, London: Overseas Development Institute (ODI).

Adams, M., S. Sibanda, and S. Turner (1999), 'Land tenure reform and rural livelihoods in southern Africa', *Natural Resource Perspectives 39*, London: ODI.

Alden Wily, L. (2000), 'Land tenure reform and the balance of power in eastern and southern Africa', *Natural Resource Perspectives 58*, London: ODI.

Anderson, Mary B. (1999), *Do No Harm: How aid can support peace, or war*, Boulder: Lynne Rienner.

Badiane, O. and M. Kherallah (1999), 'Market liberalization and the poor', *Quarterly Journal of International Agriculture*, Vol. 38, No. 4, pp. 341–58.

Baumann, P. (2000), 'Sustainable livelihoods and political capital: Arguments and evidence from decentralization and natural resource management in India', Working Paper 136, London: ODI.

Bebbington, A. (1999), 'Capitals and capabilities: A framework for analysing peasant viability, rural livelihoods and poverty', *World Development*, Vol. 27, No. 12, pp. 2021–44.

Bebbington, A. and O. Sottomayor (1998), 'Demand-led and poverty-oriented or just subcontracted and efficient? Learning from (semi-) privatized technology transfer programs in Chile', *Journal of International Development*, Vol. 10, No. 1, pp. 17–34.

Behnke, R. H., I. Scoones, and C. Kerven (1995), *Range ecology at disequilibrium: New models of natural variability and pastoral adaptation in African savannas*, London: IIED/Commonwealth Secretariat.

Berdegué, J., T. Reardon, G. Escobar, and R. Echeverría (2000), 'Policies to promote non-farm rural employment in Latin America', *Natural Resource Perspectives 55*, London: ODI.

Bradbury, M. (1998), 'Normalizing the crisis in Africa', *Journal of Humanitarian Assistance*, http://www.jpa.sps.cam.ac.uk/a/a603.tm, posted on 19 May 1998.

Bryceson, D. (2000), 'Rural Africa at the crossroads: Livelihood practices and policies', *Natural Resource Perspectives 52*, London: ODI.

Bryceson, D. and J. Howe (1993), 'Rural household transport in Africa: Reducing the burden on women?', *World Development*, Vol. 21, No. 11, pp. 1715–28.

Carney, D. (1995), 'Management and supply in agriculture and natural resources: Is decentralization the answer', *Natural Resource Perspectives 4*, London: ODI.

Carney, D. (1996), 'Formal farmers' organizations in the agricultural technology system: Current roles and future challenges', *Natural Resource Perspectives 14*, London: ODI.

Carney, D. and J. Farrington (1999), 'Natural Resources Management and Institutional Change', London: Routledge.

Christoplos, I. (2000), 'Natural Disasters, Complex Emergencies and Public Services: Rejuxtaposing the narratives after Hurricane Mitch', in Paul Collins (ed.), *Applying Public Administration in Development: Signposts for the Future*, Chichester: John Wiley and Sons.

———— (1998), 'Public Services, Complex Emergencies and the Humanitarian Imperative: Perspectives from Angola', in Minogue, Polidano, and Hulme (eds), *Beyond the New Public Management: Changing Ideas and Practices in Governance*, London: Elgar.

Clauss, B and R. Clauss (1991), *Zambian bee-keeping handbook*, Ndola, Zambia: Mission Press.

Clay E., N. Pillai, and C. Benson (1998), 'Food Aid and Food Security in the 1990s: Performance and Effectiveness', Working Paper 113, London: ODI.

Conway, G. and J. Pretty (1991), *Unwelcome Harvest: Agriculture and Pollution*, London: Earthscan.

Conyers, D. (1985), 'Decentralization: A framework for discussion', in H. A. Hye (ed.), *Decentralization, Local Government Institutions and Resource Mobilization*, Comilla: Bangladesh Academy for Rural Development, pp. 22–42.

Coulter, J., A. Goodland, A. Tallontire, and R. Stringfellow (1999), 'Marrying Farmer Cooperation and Contract Farming for Service Provision in a Liberalizing Sub-Saharan Africa', *Natural Resources Perspectives 48*, London: ODI.

Crane, E. (1990), *Bees and bee-keeping: Science, practice and world resources*, Oxford: Heinemann Newnes.

Delgado, C., M. Rosegrant, H. Steinfeld, S. Ehui, and C. Courbois (1999), 'Livestock to 2020: The next food revolution', Food, Agriculture, and the Environment Discussion Paper 28, Rome: IFAD.

Devereux, S. (2000), 'Making less last longer: Informal safety nets in Malawi', Discussion Paper 373, Sussex: IDS.

Duffield, M. (1998), 'Post-modern conflict: Warlords, post-adjustment states and private protection', *Journal of Civil Wars*, Vol. 1, No. 1, pp. 65–102.

Du Guerny, J. (1999), 'AIDS and Agriculture: Can agricultural policy make a difference?', in *Food, Nutrition and Agriculture*, No. 25, Rome: FAO.

Fichtl, R. and A. Adi (1994), *Honeybee flora of Ethiopia*, Weikersheim: Margraf.

Foster, M., A. Brown, and F. Naschold (2000), 'What's different about agricultural SWAps?', paper prepared for the DFID Natural Resource Advisers Conference 10–14 July 2000.

Gilling, J., M. Rimmer, A. Duncan, and S. Jones (2000), 'The Role of the State in Rural Poverty Reduction: Where do Sector-Wide and Sustainable Livelihoods Approaches fit in?', Background Paper prepared for the DFID Natural Resource Advisers Conference 10–14 July 2000.

Girishankar, N. (2000), 'Securing the public interest under pluralistic institutional design', in Paul Collins (ed.), *Applying Public Administration in Development: Signposts for the Future*, Chichester: John Wiley and Sons.

Gnägi, A. (1992), *Bienenhaltung im Arrondisement Ouélessébougou, Mali: Lokale Imkerei-Technologie und Möglichkeiten zur partizipativen Weiterentwicklung*, Bern: Institut für Ethnologie.

Goldman, I., James Carnegie, Moscow Marumo, David Munyoro, Elaine Kela, Somi Ntonga, Ed Mwale (2000), 'Institutional support for sustainable rural livelihoods in southern Africa: Framework and methodology', *Natural Resource Perspectives 49*, London: ODI.

Goodhand, J. (2000), 'From Holy War to opium war? A case study of the opium economy in North-eastern Afghanistan', *Disasters*, Vol. 24, No. 2, pp. 87–102.

Govereh, J., J. Nyoro, and T. S. Jayne (1999), *Smallholder commercialization, interlinked markets and food crop productivity: Cross-country evidence in east and southern Africa*, Michigan: Department of Agricultural Economics and Department of Economics, Michigan State University.

Harold, Peter and Associates (1995), 'The Broad Sector Approach to Investment Lending', World Bank Discussion Papers, Africa Technical Department Series 302.

Hobley, M. and K. Shah (1996), 'What makes a local organization robust? Evidence from India and Nepal', *Natural Resource Perspectives 11*, London: ODI.

IASC (1999), IASC Reference Group on Post-conflict Reintegration: Summary synthesis of report and field responses, 18 November 1999.

Joshi, A. and M. Moore (2000), 'The mobilising potential of anti-poverty programmes', Discussion Paper 374, Sussex: IDS.

Khan, M., D. J. Lewis, A. A. Sabri, and M. Shahabuddin (1993), 'Proshika's livestock and social forestry programmes', in Farrington and Lewis (eds), *Non-governmental Organizations and the State in Asia*, London: Routledge, pp. 59–65.

Kidd, A. D., J. Lamers, V. Hoffmann, and P. P. Ficarelli (2000), 'Privatising Agricultural Extension: caveat emptor', *Journal of Rural Studies*, Vol. 16, pp. 95–102.

Killick, T. (2000), 'Economic change and the welfare of the rural poor', Background paper prepared for the International Fund for Agricultural Development's (IFAD) Rural Poverty 2000 project.

Kortblech-Olesen, R. (2000), 'Export opportunities for organic food from developing countries', IFOAM Reports on Organic Agriculture Worldwide, May 2000.

Kydd, J., A. Dorward, and C. Poulton (2000), 'Globalization and its implications for the natural resources sector: A closer look at the role of agriculture in the global economy', Paper prepared for the DFID Natural Resource Advisers Conference 10–14 July 2000.

Lamers, J. P. A., G. Dürr, and P. Feil (2000), 'Developing a client-oriented, agricultural advisory system in Azerbaijan', TARD 2000, Rome: FAO.

Macrae, J. (2000), 'Studying up, down, and sideways: Toward a research agenda for aid operations', in L. Minear, and T. G. Weiss (eds), 'Humanitarian action: A transatlantic agenda for operations and research', Occasional Paper 39, Providence: Thomas J. Watson Institute for International Studies.

Manor, J. (1998), 'The Political Economy of Democratic Decentralization, Directions in development', Washington: World Bank.

Matin, I., D. Hulme, and S. Rutherford (1999), 'Financial services for the poor and poorest: Deepening understanding to improve provision', Finance and Development Programme Working Paper Series 9, Manchester: Institute for Development Policy, University of Manchester, UK.

Minogue, M., C. Polidano, and D. Hulme (eds) (1998), *Beyond the New Public Management: Changing Ideas and Practices in Governance*, London: Elgar.

Mosse, D., J. Farrington, and A. Rew (eds) (1998), *Development as Process: Concepts and Methods for Working with Complexity*, London: Routledge.

Mustafa, S., S. Rahman, and G. Sattar (1993), 'Bangladesh Rural Advancement Committee: Backyard poultry and landless irrigators programmes', in J. Farrington and D. J. Lewis (eds), *Non-governmental Organizations and the State in Asia*, London: Routledge, pp. 78–82.

ODI (2000), 'Can there be a global standard for social policy? The "social policy principles" as a test case', Briefing Paper, 2000(2), London: ODI.

Ostrom, E., L. Schroeder, and S. Wynne (1993), *Institutional Incentives and Sustainable Development: Infrastructure Policies in Perspective*, Boulder, Colorado: Westview Press.

Palidano, C. (1999), 'The New Public Management in Developing Countries', IDPM Public Policy and Management Working Paper No. 13, Manchester: IDPM, University of Manchester.

Persley, G. J. and M. M. Lantin (eds) (2000), *Agricultural Biotechnology and the Poor*, Washington: CGIAR.

Pinstrup-Andersen, P. and M. J. Cohen (1998), 'Aid to developing country agriculture: Investing in poverty reduction and new export opportunities', 2020 Brief No. 56, October, Washington: IFPRI.

Richards, P. (1996), *Fighting for the Rainforest: War, youth and resources in Sierra Leone*, London: Heinemann.

Rivera, W. M. and W. Zijp (2001), 'Introduction' in W. M. Rivera (ed.), *Contracting for Agricultural Extension: International Case Studies and Emerging Practices*, New York and Oxford: Oxford University Press.

Rocha and Christoplos (2000), 'Disaster Mitigation and Preparedness after Hurricane Mitch', Unpublished paper prepared for the British Red Cross Disaster Mitigation and Preparedness Project.

Rosset (1997), 'Towards an agroecological alternative for the peasantry', Institute for Food and Development Policy (Food First).

Satterthwaite, D. (2000), 'Seeking an understanding of poverty that recognizes rural–urban differences and rural–urban linkages', Unpublished paper, London: IIED.

Saxena, N. C. (2001), 'How have the poor done? Mid-term review of India's ninth five-years plan', *Natural Resource Perspectives*, Paper 66, London: Overseas Development Institute.

Schiff, M. and A. Valdes (1992), 'A Synthesis of the Economics in Developing Countries', *The Political Economy of Agricultural Pricing Policy*, Vol. 4, Baltimore: The John Hopkins Press.

Sen, A. (1999), *Development as Freedom*, New York: Oxford University Press.

Sen, A. and J. Drèze (1989), *Hunger and Public Action*, Oxford: Clarendon.

Streeten (2000), 'Globalization: threat or opportunity?', in Paul Collins (ed.), *Applying Public Administration in Development: Signposts for the Future*, Chichester: John Wiley and Sons.

Sutherland, P. D. (1998), 'Answering globalization's challenges', ODC Viewpoint, October, Washington: Overseas Development Council.

Swiss Centre for Agricultural Extension and Rural Development (2000), *Internet debate: Financing extension for agriculture and natural resource management*, Posted at *http://www.financingextension.com/*.

Tendler, J. (1997), *Good Government in the Tropics*, Baltimore: Johns Hopkins University Press.

———— (1993), 'Tales of dissemination in small-farmer agriculture', *World Development*, Vol. 21, No. 10, pp. 1567–81.

Toulmin, C. and J. F. Quan (eds) (2000), *Evolving Land Rights, Policy and Tenure in Africa*, London: DFID/IIED/NRI.

Tripp, R. (2000), 'Crop Biotechnology and the Poor: Issues in the Technology Design and Delivery', London: ODI.

Turral, H. (1998), *Hydro Logic? Reform in Water Resources Management: Lessons for Developing Nations*, London: ODI.

United Nations (1999), *World Economic and Social Survey, 1999*, New York: UN.

UNAIDS (2000), 'Report on the global HIV/AIDS epidemic', *http://www.unaids.org/epidemic_update/report/index.html*.

USAID (1996), 'AIDS Briefs', *http://www.usaid.gov/regions/afr/hhraa/aids_briefs/intro.htm*.

Wainwright, D. (2000), Paper presented at the international symposium 'Sustainable Livelihoods: Exploring the role of beekeeping in development', University of Wales Swansea, UK, 18–20 September 2000. To be published in the proceedings by Bees for Development and the Centre for Development Studies, Swansea.

Winpenny, J. (1994), *Managing water as an economic resource*, London: Routledge.

World Bank (1999), *World Development Report, 1999/2000: Entering the 21st Century*, Washington, DC: World Bank.

———— (1997), *World Development Report*, Washington, DC: World Bank.

World Trade Organization (1999), *Annual Report, 1999: International Trade Statistics*, Geneva: WTO.

# 2
# Extension, Poverty, and Vulnerability in India

*Rasheed Sulaiman V.* • *Georgina Holt*

## THE COUNTRY CONTEXT

### BASIC INDICATORS

It is estimated that one-third of the world's poor live in India: 44 per cent of Indians fall below the internationally recognized dollar a day standard, and 86 per cent earn less than US$ 2 a day (for spatial distribution of the poor in India see Map 2.1). Official data indicate that every second child is moderately or severely malnourished.

Basic socio-economic data suggest some strengthening of certain economic and social indicators, but also a number of areas for concern (see also Table 2.1). First, following a sluggish period during the post-Independence policy of heavy industrialization, per capita incomes grew at an average of 3 per cent per annum post-1975, led largely by the Green Revolution, with growth in services (information technology (IT), insurance, and banking) showing the most rapid growth in the last five years. However, such services have intrinsically low absorption of the kinds of labour that the poor can provide. Second, long term rates of growth of agricultural and foodgrain production which, despite a recent slowdown, remain above population growth rates. Third, a steady decline in the per capita availability of arable land, currently half the 1961 ratio. Fourth, some steady increase in real agricultural wage rates, but with recent evidence of a slowdown, and stagnation in the poorest states. Last, a steady increase in the ratio of food exported to food imported. They were equal in the 1960s, but exports are now four times the level of imports.

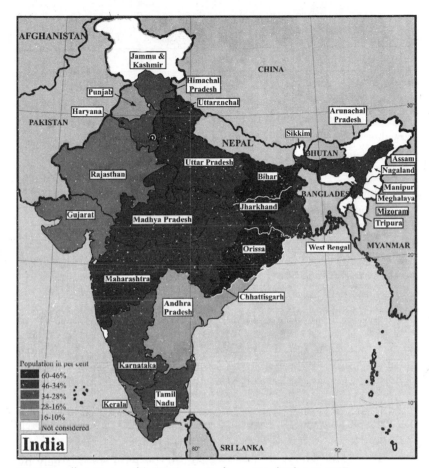

Population in per cent

- 60–46%
- 46–34%
- 34–28%
- 28–16%
- 16–10%
- Not considered

India

1992 Magellan GeographixSM Santa Barbara, CA (800) 929–4627

MAP 2.1: Spatial distribution of population below the poverty line in rural India

*Source:* Food Insecurity Atlas of Rural India (2001), M. S. Swaminathan Research Foundation (MSSRF) and World Food Programme (WFP), accessed at *http:// www.fao.org/geonetwork/images/largegifs/3744.gif*

Official statistics suggest that 26.1 per cent of the population in 1999–2000 fell below this poverty line, but more realistic estimates put this figure at around 30 per cent (Deaton and Dreze 2002). Even so, this represents a substantial reduction from 56.4 per cent in 1973–4 and 36.2 per cent in 1993–4. Around 70 per cent of the poor live in rural areas, and 70 per cent of these are primarily dependent on

TABLE 2.1
India, Basic Indicators

| Series | Value | Year |
|---|---|---|
| Cereal yield (kg per hectare) | 2339 | 2000 |
| Land use, arable land (hectares per person) | 0.16 | 1999 |
| Land use, irrigated land (% of cropland) | 35 | 1999 |
| Agriculture, value added (% of GDP) | 25 | 2000 |
| GNI per capita, Atlas method (current US $) | 450 | 2000 |
| Population, total | 1,015,923,008 | 2000 |
| Rural population (% of total population) | 72 | 2000 |
| Malnutrition prevalence, height for age (% of children under 5) | 45.5 | 1999 |
| Malnutrition prevalence, weight for age (% of children under 5) | 47 | 1999 |
| Low birthweight babies (% of births) | 34 | 1998 |
| Poverty headcount, national (% of population) | 35 | 1994 |
| Poverty headcount, rural (% of population) | 37 | 1994 |
| GINI index | 37.8 | 1997 |
| Mortality rate, infant (per 1000 live births) | 70 | 2000 |
| School enrolment, primary (% net) | 60 | 1970 |
| Surface area (sq km) | 3,287,260 | 2000 |
| Roads, total network (km) | 3,319,644 | 1999 |

*Source:* World Bank (2002).

agriculture. The largest single category of the poor are those who depend mainly on agricultural labour (approximately 40 per cent), with limited capacity of producing their own food. Malnutrition is widespread, with 207 million people in 1996–8 unable to access enough food to meet basic nutritional needs, over 50 per cent of children below 5 years underweight and girls suffering particularly badly, and anaemia prevalent among almost 50 per cent of women in the 20–49 years cohort.

Indian poverty is predominantly rural, where landless labourers and casual workers are the worst-off economic group. Scheduled castes (SCs) and scheduled tribes (STs), women and female-headed families, old people, and female children face more deprivation than others. The rural poor are primarily wage labourers and marginal farmers, that is, those with limited ownership of assets, including land. Overall, SCs and STs constitute about 25 per cent of the rural population but account for more than 42 per cent of the poor.

There are two regions of concentration of rural poverty:

1. Eastern India—East Uttar Pradesh, north Bihar, north Bengal, coastal Orissa, Assam, and Tripura—characterized by small landholdings, but high productivity through multiple cropping.

2. Central tribal India—Bundelkhand, Jharkhand, Vidarbha, Madhya Pradesh, Chattisgarh, Rajasthan, western Orissa, Telangana—risky rainfed farming supplemented by off-farm labouring.

Growth in agriculture in all areas will have major pro-poor employment effects, but patterns of investment need to match the circumstances—in the wetter areas, a focus on enhancing agricultural productivity, in the drier areas, on rehabilitation of the natural resource base (for example, though expanded and better-managed programmes of microwatershed development). Support to off-farm job creation will also be particularly important in the drier areas. In general, growth in the future must rely less on regulated markets and on subsidies on fertilizer, water, and power, and more on higher investments in irrigation, seeds, power, and roads (all of which have suffered investment declines recently).

## Policies Towards Agriculture and Rural Development

Agriculture (including cropping, animal husbandry, forestry and agroforestry, fisheries, and agro-industries) currently accounts for 26 per cent of the national gross domestic product (GDP) and provides employment to about 70 per cent of the workforce. The distribution of landholding is highly skewed; 78 per cent of farm holdings are small (less than 2 hectares) and in 1991 they commanded only 33 per cent of the total net cropped area. The average size of holding was 2.28 ha in 1970-1, but fell to 1.55 ha in 1990-1 due to a steady increase in the number of agricultural families but virtually no expansion of agricultural land.

Since independence, the main pillars of the agriculture and poverty reduction strategies of the Government of India (GoI) have been:

First, productivity-enhancing investments in agriculture. The GoI spends some US$ 110 million annually on agricultural research, with a further contribution of approximately one-third of this figure from the states. Expenditures on extension appear to be of approximately the same magnitude.

The policy approach to agriculture, particularly in the 1990s, has been to secure increased production through subsidies on inputs such as power, water, and fertilizer, and by increasing the minimum support price[1] rather than through building new capital assets in irrigation, power, and rural infrastructure, or improving the standards of maintenance of existing assets.[2] Public capital investment in agriculture has, in fact, declined from some Rs 40 billion per year at 1980–1 prices in the late 1970s to under Rs 30 billion per year in the 1990s. As well as generating environmental pressures, this has shifted the production base from low-cost regions to high cost regions, causing an increase in the cost of production, regional imbalance, the spread of rice to unsuitable areas (to the disadvantage for traditional crops such as coarse grains), and an increased burden of storage and transport of foodgrains. The equity, efficiency, and sustainability of the current approach are questionable. Agricultural markets remain highly regulated, and the management of price support, subsidized food distribution, and restrictions on the movement of 'essential commodities' have shown major weaknesses, with wheat and rice stocks exceeding 60 million tons in 2002. Moreover, deteriorating state finances have meant that subsidies have, in effect: (i) 'crowded-out' public agricultural investment in roads and irrigation and expenditure on technological upgrading; (ii) limited maintenance on canals and roads; and (iii) contributed to the low quality of rural power. These problems are particularly severe in the poorer states.

Second, price support, buffer stock, and subsidized provision of food. The Public Distribution System (PDS) and its variants focus on the subsidized distribution of basic (mainly food) commodities to some 75 million poor households through some 450,000 fair price shops nationwide. The GoI also operates a large number of centrally sponsored schemes (CSS) some of which involve payment or transfer 'in kind', for example, in the form of food for work or midday school meals. Some 12 million tonnes of food was distributed through fair price shops in 2001–2, against almost 9 million tonnes through CSS and other schemes. The PDS and associated costs of price support and of operating the Food Corporation of India (FCI) amount to around US$ 5 billion,[3] and, in aggregate the CSS currently cost around

US$ 5.5 billion. Together, the CSS and PDS/FCI amount to almost 2 per cent of GDP, or over 20 per cent of central government tax revenue. Although the PDS does reach many poor households, it does so at very high cost—the cost of acquiring, storing, and distributing Rs 1 worth of food approaches an additional Rs 2, where 'leakages' are almost 40 per cent.

Cash transfers of various kinds also form a part (currently very minor) of social protection and food security policy. These are made to 'deserving' categories of the population (such as old-age pensioners, and, in some states, widows). Further, cash is injected into certain types of activities—such as watershed rehabilitation—which although not specifically poverty focused, are geared towards some of the poorest areas, and thus, inevitably include many poor people. A number of 'hungry season' wage employment schemes also have an element of cash payment. To enhance cash transfers—and there is considerable scope for this (Farrington et al. 2003)—would increase and stabilize the demand for agricultural produce in local markets, which are often weakly integrated into larger markets.

## RURAL POVERTY: POLICIES TOWARDS AGRICULTURE AND DEVELOPMENT

A series of interventions[4] initiated in the mid-1960s, that led to the Green Revolution in cereals production, transformed the country from a situation of food deficiency to self-sufficiency. The Green Revolution was, however, restricted to productivity improvements in cereals—especially wheat and rice—in the initial decades, primarily grown in irrigated regions. In subsequent decades, productivity increased in such other crops as oilseeds, sugarcane, cotton, fruits, and vegetables. The Green Revolution generally bypassed India's vast rainfed tracts,[5] especially arid zones, hill and mountain ecosystems, and coastal regions, thus exacerbating agro-ecoregional and economic disparities (ICAR 1998). Despite its past achievements, Indian agriculture continues to face serious challenges because of its ever-increasing population, limited land and water availability, and degradation of natural resources. There are wide gaps in yield potential, and national average yields of most commodities are low.

India has one of the largest livestock populations in the world. Most of these animals are reared in sub-optimal conditions because of the low socio-economic status of their owners. The fisheries sector plays an important role in the socio-economic development of the country. It is an important source of livelihood for a large economically backward section of the population, particularly in coastal areas. Since the 1980s, growth in the production of milk, meat, poultry, and fisheries has been very rapid.

Impressive increases in agricultural production during the last four decades have improved the per capita availability of food. While extensive famines have been prevented, widespread endemic hunger still prevails among the economically underprivileged. India has made great strides in reducing its high level of poverty since the early 1970s when 55 per cent of the population was living below the poverty line, to the 36 per cent, still below the line in 1993–4, and, officially, some 26 per cent in 2001–2.[6, 7] However, the reduction in poverty levels is not fast enough to reach the Ninth Plan target of 16.5 per cent in 2001–2. Poverty in India remains predominantly rural; three out of every four poor persons live in rural areas.

Agricultural growth was a major factor in reducing poverty in India in the 1980s. Even though the rate of agricultural growth in the 1990s was similar to that in the 1980s, agricultural growth in the 1990s has had less effect on poverty reduction.[8] The PDS, Employment Guarantee Scheme (EGS) and Integrated Rural Development Programmes (IRDP) are three programmes considered central to India's strategy on poverty reduction. India is also implementing a number of nutrition programmes, such as the Integrated Child Development Service (ICDS) and school feeding programme. However, there is considerable variation in the performance of these programmes across states, mainly resulting from the varied capacity of the states to formulate and implement viable schemes (for more details, see Saxena 2001).

## AGRICULTURAL EXTENSION: BACKGROUND AND STATUS

### BACKGROUND

As in many other developing countries, extension services in India have traditionally been funded and delivered by the government. Organized attempts in this direction started after the country became independent in 1947. Pre-independence efforts were largely local attempts, driven mainly by the humanitarian efforts of a few individuals and organizations. These were area-specific and had limited impact.

Independent India acknowledged the relevance of extension quite early, a decade before organized attempts to strengthen agricultural research were initiated in the country. External aid for agricultural development emphasized extension in the 1950s. Two important programmes, the Community Development (CD) and the National Extension Service (NES) were clear examples of the commitment of the GoI to provide a number of services in such areas as agriculture, health, animal husbandry, etc. to all sections of society. With little progress on the agricultural front, the need to pay special attention to agriculture was realized, and since the 1960s many new programmes that aim to raise agricultural production have been initiated.

Until the 1960s, agricultural extension was purely a function performed under the guidance of the state departments of agriculture (DoA). A few voluntary organizations were also doing effective work in their limited areas of jurisdiction. The Indian Council of Agricultural Research (ICAR) first became involved in extension activities in 1966, with the National Demonstration Programme. The involvement of ICAR increased considerably in later years, with the initiation and spread of *krishi vigyan kendras* (farm science centres, KVK). The ICAR also initiated programmes such as the Lab-to-Land Programme and the Operational Research Programme that were merged with the KVKs in the 1990s. The establishment of radio stations and the initiation of rural programmes resulted in the wider use of the mass media for agricultural development. The print media followed suit. State agricultural universities (SAUs) initiated training programmes (for officials and farmers), demonstrations and exhibitions, and these were strengthened with the establishment of the Directorate of Extension in each SAU. Organizations created for the promotion of specific commodities (commodity boards) and specific areas (command area development authorities) also initiated extension activities. Extension was treated essentially as a public good, and with only the public sector involved with technology development and transfer, the focus was on spreading the reach of extension to all parts of the country through more extension staff and a large number of programmes.

The 1980s saw most of the states embracing the World Bank funded training and visit (T&V) system. It improved the funding and manpower intensity of extension and introduced a unified command system of extension. The studies of the T&V system that largely ignored the agro-climatic and socio-economic diversity of the country produced mixed results. A review of evaluation studies of the T&V system revealed its impressive gains (in terms of productivity) in irrigated areas and its failure to make an impact in a majority of rainfed areas. The need for a proper analysis of institutional and socio-economic factors in rainfed areas, and the importance of social science skills in making relevant interventions was also highlighted (Farrington *et al.* 1998).

Since the 1980s, more and more NGOs, agro-input industries, and agro-processors have become involved in agricultural extension activities. Now farmers' associations and producers' co-operatives are also involved in extension services for selected crops and commodities. A large number of extension services are being provided by input agencies, especially fertilizer companies. With increases in rural literacy, newspapers are devoting more space to reports related to the use of agricultural technology.

With external support drying up over the past decade, many states have found T&V unaffordable, and the 1990s saw them experimenting with the provision of alternative extension services (see Box 2.1).

---

BOX 2.1

Innovative Approaches to Extension in States

Kerala
The DoA was decentralized in 1987 through the creation of Krishi Bhavan (offices of DoA) in each *panchayat* (village council). In 1989, the group approach to extension for rice farming was adopted and subsequently extended to other crops. The European Commission (EC) funded Kerala Horticultural Development Programme (KHDP) utilizes self-help groups (SHGs), master trainers, collective marketing, and credit packages for leasing land (for more details see Isvarmurti 2000). The KHDP was initiated in 1993 and is implemented through an organization created specifically for the programme that includes consultants and 250 young graduates in agriculture, business administration, and other social sciences. The programme has recently been converted to a non-profit limited company called the Vegetable and Fruit Promotion Council, Kerala.

Rajasthan
Group approaches were adopted in the 1990s and currently village extension workers (VEWs) operate mainly through *kisan mandals*, that is, groups of 20 farmers. Under the World Bank funded Agricultural Development Project (ADP), Rajasthan encouraged NGOs to participate in extension and contracted out some services to a small number of them. In

a few blocks and districts this contracting amounts to the entire responsibility for extension. The state has also experimented with *kisan mitras* (para-extension workers).

## Uttar Pradesh

Grassroot links to extension have been weakened by the redeployment of *kisan sahayaks* (agricultural assistants) as multipurpose village panchayat development officers (VPDOs). Only 6000 of the 50,000-plus *gram* panchayats have kisan sahayaks, as VPDOs. In the rest of the gram panchayats, there is no kisan sahayak and a multipurpose VPDO drawn from another department is involved in extension work. However, the kisan mitra programme is currently operating on a large scale with the intention of covering every panchayat.

Two major programmes (both World Bank funded) are at present under implementation: the Sodic Land Reclamation Project (UPSLRP) and the Uttar Pradesh Diversified Agricultural Support Project (UPDASP). The first is committed to farmer-led extension using kisan mitras, *mahila* kisan mitras (women para-extension workers), group leaders, master trainers, and commodity-based, farmer interest groups (FIGs) for implementation at the field level. The UPSRLP is implemented through a government undertaking (Uttar Pradesh *Bhoomi Sudhar Nigam*) set up specifically for the purpose. The UPDASP encompasses human resource capacity building within line departments, decentralization of technical and managerial decision-making through the Agricultural Technology Management Agency (ATMA), the deployment of self-help groups (SHGs) and FIGs, and an increased role for the private sector. The agricultural component of the UPDASP is implemented by DoA field functionaries.

## Maharashtra

A single-window system was adopted in 1988, by merging the departments of agriculture, horticulture, and soil conservation at the operational level, which effectively improved field manpower intensity.

## Punjab

The SAU–farmer direct contact method has been used for over the past two decades, and all front-line extensionists have now been upgraded to graduate level. Punjab Agricultural University employs its own multidisciplinary extension team in each district, engaged in adaptive research, training, and consultancy.

## Andhra Pradesh

The SAU established a District Agricultural Advisory Technology Centre (DAATC) in all districts to refine technology, make diagnostic visits and organize field programmes in collaboration with the DoA and allied departments.

These experiments included decentralization (extension planning and control under elected bodies at the district or block level), contracting NGOs for some extension activities, the adoption of group approaches (instead of the earlier individual approach), the use of para-extension workers (as substitutes for DoA field extension workers) and the setting up of multi-disciplinary SAU teams at the district level. Another trend has been the formation of specific organizations (which are less bureaucratic, more flexible, and have wider expertise) to implement special programmes related to agricultural development. This has been a reflection of the increasing inability of line departments to deliver results because of their strictly enforced hierarchies, inappropriate reward structures, lack of accountability, and limited expertise. A number of examples of innovation within the public sector are detailed in Box 2.1. What is unfortunate is that very few of these have been monitored or evaluated in ways that would answer questions about their impact on whom, and under what preconditions.

AGRICULTURAL TECHNOLOGY MANAGEMENT AGENCY

The most ambitious of the post-T&V public-sector approaches in India is the World Bank-funded ATMA programme in progress in 28 districts in seven states. Of these ventures, six are at an advanced stage (Figure 2.1). The ATMA is a registered society of key stakeholders responsible for dissemination of technology at the district level. Under ATMA, day-to-day management decision-making is decentralized and farmer participation in both planning and implementation of interventions is institutionalized (for more details see ICAR 1998; MANAGE 1999a, 1999b). As a society, ATMA can receive and use project funds in such a way as to incorporate such cost-recovery mechanisms as fees-for-service. The programmes are based on the Strategic Research and Extension Plan and prepared using a participatory approach. Farm information and advisory centres (FIAC) created at the block level act as the operational arm of the ATMA. Farmer advisory committees are also constituted at the block level. These committees include all key stakeholders and farmer representatives. However, only a few of the farmer organizations constituted to date have initiated joint production and marketing activities. To identify whether ATMAs should be replicated elsewhere, the process of its implementation, as well as its outputs need to be monitored carefully.

Agricultural Technology Management Agency Governing Board

(composed of stakeholder groups and chaired by District Collector)

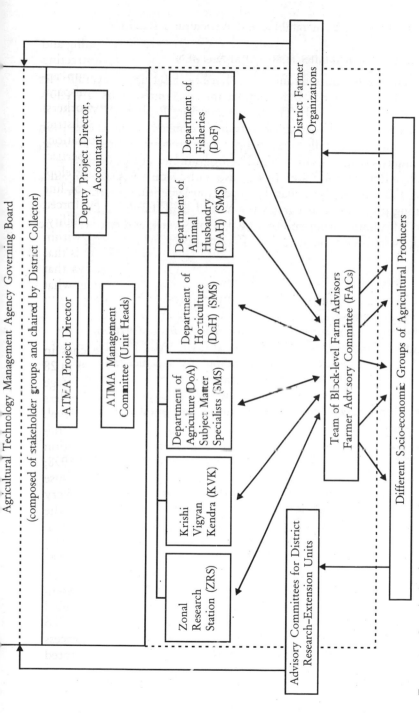

FIGURE 2.1: Organizational structure of the Agricultural Technology Management Agency

*Source:* MANAGE (1999b).

SPECIAL PROGRAMMES FOR FARM WOMEN

Women-targeted training initiatives have been set up in several states with external assistance. In Karnataka, the DoA initiated the DANIDA funded Women/Youth Training Extension Project (WYTEP) in 1983. This project has established training centres, arranged extension programmes, and collectivized input procurement, shared learning, and micro-credit activities. Similar programmes are in operation in Madhya Pradesh and Tamil Nadu (also with Danish Assistance). Since 1992, the DoA of Rajasthan has been implementing a farm women training programme as part of the World Bank funded ADP. Three types of training programmes (one-day, one-week, and two-week) are used to train farm women in different aspects of agriculture.

OTHER INITIATIVES IN THE PUBLIC SECTOR

The National Ministry of Agriculture (MoA) has initiated some important innovations during the last 20 years. Among these, the establishment of an autonomous training institute, the National Institute of Agricultural Extension Management (MANAGE) at Hyderabad in 1987 was significant. Other important initiatives include the agricultural extension through voluntary organizations scheme (initiated in 1994–5) and more recently, 'macro-management' initiatives in which 27 centrally sponsored schemes were merged thus enabling states to increase the flexibility of their programme planning and prioritization. Under a new scheme, the Central Government is providing a 25 per cent subsidy to set up 5000 agri-clinics using unemployed agricultural graduates to provide testing facilities, diagnostic and control services, and other consultancies on a fee-for-service basis. The programme is implemented through the Small Farmers Agri-Business Consortium (SFAC), and is financed principally through banks.

The national research organization (ICAR) now has 261 KVKs (—138 in SAUs, 26 within ICAR institutes, and 90 in NGOs and government bodies) organizing vocational training for farmers (Das and Hansra 1999). The Council has also strengthened 53 zonal agricultural research stations (ZARS) to assume additional functions of the work of KVKs (ICAR 2001). Other programmes of the ICAR include 8 trainer training centres, 70 institute village linkage programme (IVLP) centres, and 60 centres of technology evaluation and impact assessment. Under the National Agricultural Technology Project (NATP), the ICAR is at present establishing 40 agricultural technology information centres (ATICs), 25 in SAUs, and 15 in ICAR institutions. Through the ATICs farmers can get many services at the same place, that is, through a 'single window'; including the delivery of research products, information, and other services.

## PUBLIC AND PRIVATE SECTOR ROLES

India is actively considering various options for limiting public sector involvement in extension and is contemplating steps to complement, supplement, and replace some of its activities by greater involvement of the private sector, both commercial and non-profit (DAC 2000). However, there are concerns over public and private roles (Box 2.2). The private sector itself exhibits diverse approaches to agricultural extension (Sulaiman and Sadamate 2000; Chandra Shekhara 2001). Farmers' associations and producer co-operatives are at present involved in extension for selected crops and commodities.[10] Milk

---

### BOX 2.2
### Public–Private Partnership

There is an increasing realization that, 'Public extension by itself cannot meet the specific needs of various regions and different classes of farmers. Policy environment will promote competitive, private and community extension to operate effectively, in roles that complement, supplement, work in partnership, and even substitute for public extension' (DAC 2000). Though this declaration seems to portray a genuine response to the changing times, the level of preparedness in the public sector to work in a multi-institutional environment is not encouraging. The GO–NGO collaboration experiment for sustainable agricultural development implemented in Rajasthan highlights a number of issues that emerged when the government and NGOs were brought together to work in a collaborative mode, including the pervasive perception in the government that NGOs should merely be contracted to provide services, but a perception among NGOs that their strength lies more in mobilizing people to make demands on the system (Aslop *et al.* 1999).

Private sector involvement in extension is not uniform across the country. In some districts, a number of private organizations work in isolation providing diverse services, but there is as yet no inventory of these that would allow contact to be made and possible alliances forged. As long as the major responsibility of the DoA continues to be the implementation of schemes, they will see no reason to link with the private sector. Neither does the past experience of research-extension (R–E) linkage inspire confidence in the partnership approach. Partnerships, by definition, require sustained efforts over time to build and maintain good working relationships. As a first step, a series of activities to inculcate the right attitude for building close working relationships with other organizations needs to be arranged for extension managers. Case studies based on real documentation of successful partnerships disseminated between regions would also act as a springboard for discussion around the concept, operation, pitfalls, and potential of partnerships.

Use of the term 'privatization' has created confusion among public extension personnel. The draft policy framework for agricultural extension prepared by the DAC calls for increased participation of the private sector in agricultural extension, but the response of state governments has not been encouraging. Many states have expressed strong reservations about the idea of private participation in extension due to the profit motive of the private sector, the lack of ability of farmers to pay, the lack of an effective private sector, and the need to preserve the 'authenticity' of agricultural research derivable only from an 'impartial' public sector.[9] The fact that privatization is an umbrella term covering several options by which to improve efficiency and effectiveness has not been fully appreciated, and awareness of the mixed success that has been reported for privatization strategies internationally, further impedes the adoption of such a strategy.

Although private sector participation in extension in India is currently limited to only a few crops and geographical areas, an increasing number of private entities, such as NGOs, farmers' associations, producer co-operatives, input agencies, agro-processors (especially for contract growing schemes), private consultants and the media offer much scope for supplementing and complementing public sector extension. As farmers are also willing to pay for value-added services, the challenge is to create quality services so that cost recovery can commence. In an increasingly complex environment, there are a number of responsibilities, such as certain designated public goods, which remain the domain of the public sector, though in reality and as suggested in the literature the boundary between public and private goods is not so clear. For these services to be funded primarily from the public purse does not necessarily require them to also be delivered through public infrastructure. The more the institutional pluralism grows, the greater the need within the public sector for clear mission statements, goals and strategies so that privatization options are used to best advantage.

co-operatives in various states provide a number of services to dairy farmers. Other important providers include NGOs, input industries (seed, fertilizer, pesticides, etc.) and agro-processors servicing contract farmers, such as Pepsico in Punjab; Vazir Sultan Tobacco Company Limited (VST), Natural Products in Andhra Pradesh; media organizations, such as E-TV in Andhra Pradesh and Maharashtra; print media in Kerala; and private consultants (Box 2.3).

---

Box 2.3

Examples of Private Sector Extension in India

Maharashtra Grape Growers' Association (MGGA) is one of the best-known farmers' associations in the country. With an estimated membership of 17,000 growers, MGGA has been the driving force behind the development of grape cultivation in the state. The Association has an independent research and development (R&D) wing, organizes regular group discussions and seminars and publishes a monthly farm magazine on grape farming.

Malabar Regional Co-operative Milk Producers' Union (MRCMPU) is a part of the Kerala Co-operative Milk Marketing Federation, and operates through 429 producer societies. Its extension cell organizes technical inputs, training and extension, artificial insemination and veterinary services. Under the Women Cattle Care programme, knowledge is transferred through women who act as 'village change agents'. They conduct regular informal discussions and organize formal classes for groups of 20–5 women.

Indian Farmers' Fertilizer Co-operative Limited (IFFCO), Fertilizer and Chemicals Travancore Ltd (FACT) and Krishak Bharati Co-operative (KRIBHCO) are some of the more prominent fertilizer organizations involved in providing demonstrations, village adoption programmes, farmer visits to research stations, and soil testing services.

Kumar Gentech and Tissue Culture Co., Pune provides total extension support from advice on site selection, to technological guidance throughout the growing period and marketing support for those growers buying inputs (seeds) from the company. Agro-processors such as Pepsico provide total extension support to contract growers of tomato in Punjab.

Rallis India Limited, primarily a pesticide manufacturing company and Mahindra Industries, primarily a tractor manufacturing firm, are experimenting with strategies to provide integrated yield and profitability improvement solutions to farmers through knowledge, access to good quality inputs, farm mechanization services, finance, and marketing access. To facilitate farmer interaction and provide these services, they establish centres in the selected districts.

Samaikya Agri-tech Private Limited is a consultancy firm operating in Andhra Pradesh which provides extension services to farmers. It advises farmers registered with the firm on all aspects of farming, ranging from crop production to diversification and marketing options. Subject matter specialists of the firm at the district level provide technological back-up to the technical officers in the field through online connectivity.

PAN Horti Consultants and Viji Hi-Tech are two consultancy firms in Coimbatore district, Tamil Nadu providing consultancy services on agriculture to commercial firms, agro-based industries, and entrepreneurial farmers.

In Kerala, 18 Malayalam (local language) dailies publish regular agricultural columns every week. Farm magazines published by several media firms in the local languages of different states, provide useful information to farmers. Private TV channels such as E-TV transmit daily agricultural programmes in Telugu and Marathi.

MS Swaminathan Research Foundation implements the information village concept in villages around Pondicherry. Information support on all aspects including agriculture is provided by villagers trained in information technology.

EID–Parry, an established sugar company in South India provides contract growers access to websites on agriculture in the company's field offices. Apart from information on sugarcane cultivation, the site provides useful information in the local languages on other crops, new technologies, and prevailing market prices.

India has a number of NGOs with varying levels of capacity, implementing a wide range of programmes in several States. They include: Bharatiya Agro-Industries Federation (BAIF), Professional Assistance for Development Action (PRADAN), and Action for Food Production (AFPRO).

Despite perennial weaknesses (that is, diminishing operational support and poor technical background of the majority of its employees), the VEW of the DoA is still the most important source of information for farmers in India. This is despite the fact that the information is narrowly targeted to grain production, visits are irregular, and the service is preoccupied with the implementation of public sector schemes linked to subsidies and subsidized inputs. In the more remote and difficult areas, the DoA has considerable difficulty in recruiting and retaining field staff and, these areas, often have large numbers of vacancies and frequent staff turnover (Box 2.4). The main extension function performed by the state DoA is the delivery of

---

Box 2.4

### Chronic Difficulties of Publicly Provided Extension in Remote Areas

The majority of the rural poor in India live in areas weakly integrated into markets. Apart from a few NGO initiatives, the majority of private sector innovations do not reach these areas, and so the only available extension service is that provided by the state government. This service is largely dysfunctional because of three types of chronic difficulties:

• All government extension workers are permanent and pensionable civil servants, their accountability to clients is limited, and promotion depends more on the number of years in a post and than on capability. Most staff consider remote areas to be 'punishment postings' and many, newly recruited into these areas, spend a large part of their time seeking transfers to more-favoured locations. As observed in remote parts of Udaipur district in Rajasthan (Alsop *et al.* 1999), on an average, almost 50 per cent of the posts are vacant.

• Efforts to 'broad-base' extension are in principle undoubtedly sound, that is, extension workers should be able to advise on agriculture in its broad definition, and not just on crops, but also on aspects of input supply, processing, and marketing and increasingly, on the implications of new market specifications for production and processing technology. However, the capabilities of those willing to live in remote areas are usually limited, and improved impact across such a broad canvas would require long-term re-training (which is rarely available) and more impact-oriented reward structures, which are unlikely to be introduced within a permanent civil service structure.

• Middle-management tends to be preoccupied with meeting targets that are inflexibly interpreted, lack client-orientation, and are uninformed by any kind of institutional learning. As a consequence, field-level agents lack the space to try new ways of meeting client needs.

These three factors suggest that publicly funded, publicly implemented services in the more remote areas of India face chronic difficulties and without considerable reform are unlikely to serve as a basis for greater poverty orientation in extension.

technical messages to individual farmers or farmer groups through visits to specific locations in a particular circle or area. The extent of satisfaction with the information support provided by DoA varies widely. Farmers' dependence on other farmers and input dealers as a source of information continues to be high.

Farmers' associations and producers' co-operatives provide a large number of services, including extension, to farmers but they exist only for a few crops or commodities and in few locations. The same is true of commodity boards. The field extension activities of SAUs and ICAR research stations are restricted to a few villages in their immediate vicinity. The KVKs have a number of vocational training programmes for farmers. With very few exceptions, most of the NGOs are small and their activities, though intensive, are restricted to small numbers of beneficiary farmers in a limited number of villages. Consultancy services are few and are mostly private ventures often associated with high-value crops. However, the number of organizations providing integrated yield and profitability solutions to farmers (input, hiring machinery, consultancy, and marketing) has been increasing. The potential of the mass media is underutilized at present, but the agricultural programmes of some of the private television channels and print media provide sources of information that have high impact on commercial farming. Input companies do not have full-time extension staff. The demonstrations and seminars provided by input companies, often in collaboration with the DoA, are essentially about marketing rather than educational activities and suppliers rarely provide continued (post-purchase) support to farmers.

With the increasing realization that knowledge is an important input for efficient farming, the institutional diversity in the provision of extension services will increase in the years to come. Privately provided agricultural extension is concentrated in areas of commercial agriculture, and in the very limited areas in which NGOs operate. However, there are several districts where a number of organizations provide diverse extension services, but work in isolation from each other. Certainly, there is little motivation for public extension services to collaborate in any substantive way with others, and collaborative experiments have often floundered because of public sector tendencies to dominate the agenda.

The policy discussion document released by the DAC in late 2000 envisaged a number of significant changes in the provision of publicly funded extension in India. It must be noted that the primary responsibility to provide extension lies with the individual states, not with Central Government, and so much will depend on the degree of acceptance of these ideas by the states.[11] The fact that Central Government support is envisaged for certain types of change will undoubtedly enhance the prospects of ideas being implemented. Nevertheless, implementation is likely to be slow and uneven, and there is as yet little recognition that a number of the proposed ideas have already been tried with little success, and that others have limited prospects.[12] Nevertheless, this represents an effort to place extension debates into a much wider policy context, and so merits close consideration. The main provisions of the document are detailed in Box 2.5.

---

Box 2.5

Main Changes Envisaged in Agricultural Extension
Provided by Central Government

At the policy level
- A move towards a farming systems approach.
- Partnerships with private and other public agencies in extension provision, including:

  (a) public funding of private provision;

  (b) cost recovery for some services;

  (c) skill enhancement among farmers;

  (d) linking of technology advice to new market opportunities;

  (e) local-level accountability of extension workers to farmers.

Institutional restructuring
- Some reduction in the number of village-level extension workers, and instead a focus on small block towns (a block being an administrative unit of some 70 villages) where single-window extension services will be provided, using the ATMA model.

- Using participatory strategic research and extension plans (SREPs) to drive local-level technology generation.

- Extension delivery at the block town level complemented by strengthened farmer-interest groups capable of creating 'demand pull' on the system.

Financial reforms
- Central Government will contribute towards operation and management costs in future, though salary costs will remain the responsibility of the states.

---

- Additional public funds will go into a number of new areas, including the payment of honoraria for para-extension workers, and support to NGOs involved in local-level group formation.

Strengthening research–extension linkages
- Preparation of SREPs, with efforts to reactivate existing interactions, such as biannual meetings between state DoAs and the SAUs, and the national pre-season meeting between ICAR and the DAC.

Capacity building and skills upgrading
- Central Government will support training for extensionists once the states have formulated a human resource development policy for extension.
- Such training will include social science and IT components, previously not incorporated into training.
- All agencies (public and private) will be networked electronically to state headquarters, the SAUs, and MANAGE.

Mainstreaming women in agriculture
- Women's access to extension and training will be enhanced.
- Male extension workers will be sensitized to the needs of women farmers.
- Civil service rules will be examined for gender bias.
- Access by female extensionists to training will be improved.

Use of media and information technology
- Provision of online market information.
- Support to the private sector to establish IT information kiosks.
- Wider use of mass-media for extension.
- More farmer participation in mass-media programmes.

Financial sustainability
- Provisions to privatize the 'private goods' elements of extension, especially in more favoured areas.
- Provisions for cost recovery.
- Co-financing of extension via farmers' organizations.
- Liberalization of the regulations governing commercial activities by training centres to allow profits to be retained.

Changing role of government:

- The role of government is seen largely in the neo-liberal terms of provision of public goods, and the creation of an enabling environment for efficient functioning of the private sector, with separate provision to make good any market failures not otherwise addressed.

*Source:* DAC (2000).

THE EXTENSION ROLE OF RESEARCH ORGANIZATIONS

State agricultural universities and ICAR institutes are engaged in some limited extension activities. The SAUs have a directorate of extension involved in training, publicity, and advisory services. The SAUs in Punjab and Andhra Pradesh have created mechanisms for one-to-one interaction with farmers at the district level, to primarily provide problem-solving advice. The ICAR institutes provide training on technologies developed by the institute, mainly to officers of the line departments and also to farmers. The relatively new IVLP is a vehicle for technology assessment and refinement (TAR) at the village level by a multi-disciplinary team of scientists.

The increasing involvement of ICAR in extension activities has been of concern to the Council for quite some time. Questions have been raised as to the increasing share of the budget allocated to extension, and the logic of continuing full funding for KVKs. The role of social scientists in general and extension scientists in particular within research institutes, therefore, needs clearer identification, especially considering that their potential usefulness in technology development is not being realized at present. Perhaps extension scientists could be better employed in developing new institutional arrangements and linkages within the research–extension (R-E) system, and partnerships with outside organizations. To remain relevant, extension in research institutes should broaden their role from promotion of technologies to facilitation of technology development (contributing crucial social science perspectives) and application (experimenting with institutional innovations and facilitating linkages within the innovation system).

## MAKING AGRICULTURAL EXTENSION AND RURAL DEVELOPMENT PRO-POOR: OPPORTUNITIES AND CONSTRAINTS

### OPPORTUNITIES

Coutts (1995) notes that definitions of extension range from that of a pervasive technology transfer model to a facilitative human development model. Between these extremes lie other models including those of extension as an advisory or consultancy (or problems-solving) function, and as adult education. A number of developments, namely tighter government finances and economic reform policies (liberalization, redefining the role of state and private sector) have changed the way that governments fund and deliver agricultural development. Challenges on the sustainability (depleting natural resources) and trade fronts, World Trade Organization (WTO) driven changes, the changing nature of agricultural technology (from public to private goods), rapid developments in IT and a changing development agenda (stakeholder participation, decentralization and faster reduction of poverty) have prompted a re-evaluation of the role of extension in many countries, including, to a limited extent, India.

Public expenditure on agricultural extension and its control has been justified on the basis that support for agriculture leads to reduced food prices and increased food security which benefit the whole population (van den Ban 2000). More recently, additional rationales for public extension include poverty alleviation, employment creation, and environmental conservation, although these have not been explicitly mentioned in Indian policy documents. In a climate of market liberalization for the maintenance of costly public extension systems, there are important questions about who should fund and deliver extension in relation to each of these purposes. In a multi-institutional environment, it would be efficient for various actors to prioritize their activities based on their inherent strengths and weaknesses. When a market for skilled and specific agricultural advice is developing, the government should reconsider its role in this market and evaluate its comparative advantage. It is normally sensible for a government to create conditions in which private suppliers of advice can emerge and flourish. This view has been supported by others who have found merit in limiting the government's role to only those activities that are not provided by the private sector. Moris (1991) argues that the government must reduce services to levels that it can adequately fund, while supporting the private sector in the provision of the remaining services.

One approach used to decide who should provide what services, is based on the classification of services according to economic characteristics, using the principles of public and private goods.[13] Private firms are unwilling to supply services with public good characteristics because it is impossible to restrict the benefits to people who pay for those services (the free-rider problem). Thus, private enterprise will be willing to supply any good or service that can be sold for profit. The implication is that the public sector should focus its funds on the public good components of extension. In turn, publicly funded extension can either be supplied by the public sector, contracted out to private (commercial or non-profit) organizations, or delivered through public–private partnerships of various kinds. Van den Ban (2000) argues that whether the extension function is public or private depends on the context in which it is used. In theory, advising farmers on the optimal quantity of fertilizer is site and farm-specific and, therefore, private and chargeable. But, training farmers in the use of soil-testing equipment is an educational function for which public funding may be justified. This difference is critical, because by stepping back from the problem to examine its underlying cause, not only does the nature of the good change, but the value of the information also increases to the extent that it becomes empowering.

Although the private enterprise may be willing to supply a good or service, it is in many cases unable to step into the role of provider either because the environment (in terms of infrastructure and information) needed for a business to function is lacking, or the market is dominated by monopolistic interests, or the extreme levels of risk involved (as in the case of insurance services). Consequently, governments are also faced with the task of creating a suitable environment for commerce, which further diverts funds away from direct support to farmers. The net result is that the public sector takes on a complex of different roles ranging from provider to co-ordinator, facilitator, arbitrator, regulator, and guarantor (of transparency). As already argued, efforts to maintain permanent public officers at the village level in remote and difficult areas face chronic difficulty. Bearing this difficulty in mind, the government's ability to deliver quality public goods through its own staff needs to be evaluated. While the private sector could not be expected to fund these types of services, it can certainly be contracted to provide them. In the Indian context, this would help the government in keeping down the number of public sector staff, and in wielding more flexibility in staff deployment.

Some government policies towards weakly integrated areas in India amount to unmanaged forms of triage. This implies the neglect of such areas relative to others, and as a result, the population might move out spontaneously to seek livelihoods elsewhere.[14] A classic example of focusing resources on well-integrated areas was that of the Intensive Agricultural District Programme (IADP) in 18 districts in 1960, when India was reeling from food shortages, and it was essential to have quick results on the production front.[15] But in many cases, where triage does exist, it is more a product of poor implementation rather than that of policy design deliberately geared against areas weakly integrated into markets. States, regions, and districts that are relatively less-developed (for example, rainfed), remote (hill and desert areas) and have a high proportion of the population belonging to weaker sections (tribal people) have recently been the focus of specific government plans, such as the panchayat extension to scheduled areas, and programmes for drought-prone areas, deserts, and watershed development. The implementation of these schemes, however, remains weak mainly due to the lack of participation in selection and implementation of programmes, inflexible implementation guidelines, lack of voice among the poor, and inadequate personnel to implement programmes. What is needed for the future is recognition that:

1. there will be outmigration from the more remote areas, and so there is a need to support migrants, and to regulate the pace of migration to be consistent with absorptive capacities;

2. there are agriculture-related possibilities, and that these are not driven by productivity enhancement alone, but also by requirements to reduce vulnerability and generate employment opportunities; and

3. agriculture alone is unlikely to be sufficient for the livelihoods of the poor—support is also be needed for non-farm enterprises, and improvement of access by poor people to services and employment opportunities in nearby towns.

Farmers need to be supported with information, knowledge, and the skills to adopt improved technologies that result in improved farming, that are productivity enhancing, vulnerability reducing, and employment creating. However, the requirements of farmers and rural families go beyond agricultural production technologies. Changes in recent years, not least the increasing penetration of markets into rural areas and the need to tailor products to ever more stringent market requirements means that extension support must now address a broader range of farmer requirements (Box 2.6).

Against this context, publicly funded extension in India in the 1990s took on board new goals of natural resource management (especially watershed management, participatory irrigation management, and others) and diversification. The need for a group approach to extension and the importance of producer groups (farmer interest groups, commodity associations etc.) were also recognized, but many of these changes remained at the level of planning and rhetoric. For reasons relating to the characteristics of public sector extension

Box 2.6

Potential Forms of Extension Support in India

- choice of technological options appropriate to available land, capital, labour, and knowledge resources;
- management of technologies, such as the optimal use of new inputs;
- decisions about how and when to change enterprise or farming system, such as diversifying from crop production to mixed farming or vegetable or animal production;
- assessing both domestic and foreign market demands for products and product quality criteria within these markets, for example food safety and organic criteria;
- sourcing reputable suppliers of inputs and forging trust-based alliances with them;
- co-operation between small-scale producers to increase their presence and power in the market;
- sourcing readily accessible and accurate information on an ongoing basis;
- assessing the feasibility of off-farm and non-farm income generation opportunities to provide long-term benefits;
- assessing the implications of farm enterprise in relation to changing policies on input subsidies and trade liberalization.

*Source:* van den Ban (1998).

services, already outlined, success at the operational level has been limited. Reducing poverty, or the vulnerability of the poor, has never been explicitly stated as a goal of extension, and there have been no specific extension programmes to target the poor (although in the case of distribution of subsidies and subsidized inputs, programmes often target small-scale and marginal farmers). But with the overall policy focus shifting to the reduction of poverty levels, extension can no longer afford to ignore the poor.

A focus purely on increased food production is not enough to solve the problem. Several districts in India with high agricultural productivity levels also have high levels of poverty (Haque 2000). Food assistance programmes, though important in times of crisis, do not address the basic cause of the problem and are expensive to implement.[16] In many areas, there are limits to achievable increases in productivity unless appropriate institutions—that can help farmers to access information, inputs, and services—are strengthened, and joint action for natural resource management, marketing, and processing are promoted. Often, other livelihood options need to be explored to improve income (and thereby access to food) and extension can and should meaningfully contribute to attaining these goals. In some contexts, cash transfers as a form of social protection (for example as old age pensions) can strengthen access of the poor to food, and enhance local demand (Farrington *et al.* 2003).

The poor face substantial transaction costs in accessing the means of production, in adding value to their produce, and in accessing markets for it. The core of extension is to help people make better choices through the supply of information and to enhance their capacities to process such information and act on it, thereby reducing the transaction costs involved in pursuing livelihood options. But transaction costs impact disproportionately on the poor, due to access problems caused by weak infrastructure, poor organizations, and adverse local power relations (Christoplos *et al.* 2000). The intensity of these problems varies widely and so there cannot be a single blueprint. Moreover, cultural contexts vary significantly, thus, approaches to reducing vulnerability vary from district to district and often within the same district. Solutions for reducing poverty rarely lie in the transfer of production technologies alone, but often also in improved access to information on wider livelihood choices and institutional support (such as micro-finance, micro-enterprises, entrepreneurship development, market access, etc.). Extension needs a higher level of flexibility and wider range of expertise to choose and assist the poor with these various options. The scope for adapting IT applications to the needs of extension has received particular attention (Box 2.7).

CONSTRAINTS

Extension in India has moved towards plurality in provision. As a part of this process, it is essential that the government show willingness to systematically document public and private sector institutional innovations and drawing lessons for the replication of best practices. Currently this policy is sadly missing. A number of generic issues still plague many experiments in institutional innovation and these need to be addressed before replication on a wider scale. The key issues in the Indian context are discussed in the following sections.

Box 2.7

Information Technology and Extension

Information technology has much to contribute to improving the efficiency and effectiveness of extension systems. The spread of internet access has created considerable interest in 'cyber extension' (for details of the possibilities and potential see Sharma 2000). The excitement generated by IT has tempted many to consider opening more information kiosks (or IT parlours), and to develop online connectivity to ICAR and SAUs, to compensate for the weaknesses of public extension. But, it should be noted that IT is a useful complement to, but not a substitute for, field extension. Research has clearly shown that information supplied through the media is helpful only in the initial stages of technology adoption and that a more detailed interaction is required in the later stages of the adoption process (van den Ban and Hawkins 1996). Furthermore, poor infrastructure, high levels of illiteracy, and absence of software in local languages constrains the usefulness of IT for vulnerable farmers. Human-mediated computer systems shared among multiple users of a rural community could in fact prove to be the most inexpensive and inclusive form of rural infrastructure possible today (Sood 2001). Information technology also has a great potential to improve the quality and competence of the R–E system and to provide greater access to market information (through networking of markets) and far more attention needs to be given to this.

## Scale and Complexity

The Indian extension system has to cater to the needs of about 90 million farm holdings, 70 per cent of which belong to small-scale and marginal farmers. Public-sector extension employs about 100,000 workers. The ratio of extension workers to farm families varies from 1:300 farm families in Kerala to 1:2000 in Rajasthan. Taking account of non-public providers does not change these ratios appreciably. The DoA continues to be the only agency with a presence in all the districts. Even though it is not necessarily the primary source of information for farmers, with continued reliance on individual or group contact and little multiplication of impact through mass media, information coverage remains unsatisfactory. In remote and difficult areas, presence of both public and private sectors is weak. While the public sector has difficulty in retaining field staff, the private sector finds these areas unattractive for investment.

Given the wide variation of agricultural enterprise, productivity levels, non-farm opportunities, infrastructure, and poverty levels across and within states and districts, the current uniform approach to public extension within a state or a district is inappropriate. District extension managers must be given greater flexibility to design and use different extension approaches, and the freedom to identify strategies that could reduce the vulnerability faced by the poor. They also require more flexibility in deploying funds and staff. At the moment, such flexibility is lacking as the overriding philosophy of government continues to be strict adherence to uniform prescribed guidelines. The system of reporting and lack of incentives for experimentation further discourages any institutional innovation.

To make effective decisions, farmers need information from a wide range of sources, on farm and non-farm topics. But the public sector continues to provide information only on technologies generated in public-sector research stations and mostly on food grain crops. Efforts to respond more adequately to the diverse needs of small-scale and marginal farmers remain weak.

*Linkages*

After two decades of efforts to foster R–E linkages, information flow is still mostly top–down (Macklin 1992) with feedback too weak to catalyze the fundamental changes required in the prioritization of on-station research (Jha and Kandaswamy 1994). The DAC has recently devised fresh guidelines for establishing R–E linkages under the innovations in technology dissemination (ITD) component of NATP. But even these are a product of the linear, mechanistic model of innovation that has outlived its utility and are unlikely to change the situation on the ground, especially for the vulnerable. Understanding the need for a holistic system, and of actor-oriented approach to innovation continues to elude policy makers (for discussions of the linear *vs* systems models of agricultural innovation see Biggs 1990).

Even within the R–E system, linkages between organizations working in the same subject area are weak and this severely constrains the performance of the system as demonstrated by recent case studies from the Indian horticultural research systems (discussed in detail in Hall *et al.* 2001). Inter and intra-departmental co-ordination for programmes in both ICAR and SAUs are weak. (ICAR 1996). Linkage between KVKs and state DoAs is less than satisfactory and the DoA continues to ignore other organizations that have entered the extension arena in selected regions and enterprises that could complement or supplement its efforts.

Linkages between public sector extension and institutions whose polices have a direct bearing on extension—input supply, credit, and marketing systems—are virtually non-existent. Nor can public extension influence policy on investment, research prioritization, infrastructure, public administration, or technical education. Public extension thus continues to be a passive recipient and often a victim of decisions taken in these systems.

### Fiscal Sustainability and Operational Accountability

Inadequate operational funds have been a perennial weakness of public sector extension. Macklin (1992) noted that the level of operational funding in the T&V system had not been maintained in real terms, thus reducing the mobility of extension workers (Macklin 1992). Swanson (1996) estimated operational expenditure in state DoAs at around 15 per cent of total expenditure; considerably less than the level considered necessary for a fully functional extension system.[17] More recently, a study across four states revealed that salaries alone account for 85–97 per cent of government expenditure on the state DoA (Sulaiman and Sadamate 2000). This has resulted in serious underutilization of facilities and personnel. Its origins lie in the fact that, under Indian civil service codes, budget cuts cannot be translated into redundancies, and thus are borne entirely by operating budgets.

The problem is a vicious circle of fiscal difficulty, curtailed services, inefficient operation, depressed performance, staff demotivation, and reduced competence (Ameur 1994). Increasing budgetary constraints within states have had a knock-on effect on central DoA budgets and questions are being raised as to the financial sustainability of the vast extension infrastructure in India (DAC 2000). Measures to recover at least a part of operational expenditures are slow to be adopted because—despite some indications of willingness to pay for services (48 per cent of farmers are reportedly willing to pay for extension in horticultural and high-value crops, Sulaiman and Jha 2000)—the quality of available services is too low to meet demand. Currently, political sensitivity surrounding the issue prevents the introduction of measures that could capitalize on farmers' willingness to pay, and that would otherwise enable a cost-recovery strategy to gain momentum.

At present there are no effective mechanisms to achieve accountability in the public sector since it is subject to the rules and regulations of the civil service. Consequently, there is very little scope for applying a performance-related system of incentives. Evaluation remains subjective and there is an urgent need to evolve objective performance indicators and transparent means for implementing them.

There needs to be a reduction in the number of central sector programmes that are often implemented under uniform operational guidelines. This issue is now receiving attention and the central government is implementing a macro-management approach, which allows states to select programmes for implementation. It would be ideal if states were to also start adopting this approach for the districts.

## Human Resource Capacity

Several training organizations with varying resources and professional capabilities exist in India, but the performance of public sector extension agencies in integrating knowledge and skills from these various sources has been limited. The ability of extension agencies to assess and refine their knowledge base is poor and this is largely due to the low level of qualifications required of their employees.

There is now a fairly broad consensus that poorly trained VEWs would not be able to deliver appropriate standards of service in the changing extension environment. In total, states employ some 100,000 extension staff of whom only around 20 per cent are graduates. Some states have already taken the decision that degree level should be the minimum qualification for appointment in the DoA. But this is not enough. Department of agriculture employees need skills in rural management, social mobilization, training, finance, credit, and marketing, in addition to agricultural science.

Though enforcing graduation in agriculture as a minimum qualification at the entry level is to be welcomed, the fact remains that agricultural graduates also lack many of the social science skills necessary for field extension work.[18] Postgraduate training in agricultural extension does not address many of these skills (Sulaiman and van den Ban 2000a). As extension performance depends considerably on the quality of agricultural graduates, the extension system should find ways to address the content and quality of agricultural education.

A universal ban on recruitment in the VEW cadre, before alternatives have been adequately tested (such as efforts to base extension in block towns and stimulate farmers' capacity to draw on such services) would be premature. Experiments in introducing para-extensionist workers also need to be handled carefully: experience from Uttar Pradesh suggests that the kisan mitras, (who are selected from practising farmers) after a single-day training programme before each season are unable to provide any better information/advice to farmers. Clearly, any para-extensionist scheme would have to be better designed and implemented than this, and the ways in which it would draw on public extension resources would have to be clearly defined.

Several states also lack competent subject matter specialists (SMSs) at the district level, a major factor that contributes to poor R–E linkages and lack of integration across crop and livestock systems. Such constraints severely limit the capacity of the technology dissemination system to assist farmers in using improved production practices or incorporating higher-value commodities into their farming systems (Sharma 2000).

Many of the questions on capability and future preparedness of different extension organizations depend on a clear definition of the roles that extension organizations will play. A comprehensive evaluation of the extension capabilities available within different organizations in the public and private sectors needs to be made. Public sector infrastructure and expertise could be productively used to enhance the capability of private extension providers. Similarly, public sector extension can gain expertise on private sector technologies and products, if partnerships can be established between these different agencies. Capacity building in extension needs to make an inclusive analysis of all possible organizations so as to improve the collective performance of the extension system.

In order to provide problem-solving consultancy and initiate measures to mobilize farmers, the public extension system needs knowledge and competence in a number of areas (Figure 2.2). To remain relevant and useful in the years to come, public sector extension has to strengthen its understanding of technology, markets, prices, demand, and policies; either by recruiting specialists or by contracting out these services. The latter option might be preferable because it would ensure optimum use and accountability. Whatever public/private division of responsibilities is chosen, a sound human resource management strategy is required. This needs to be based on a clear analysis of present and future knowledge and skill requirements, the roles and responsibilities for individuals and teams,

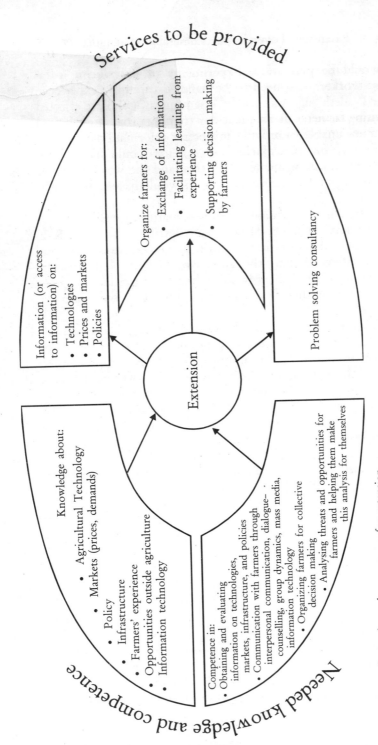

FIGURE 2.2: Future roles and competence for extension
*Source:* Sulaiman and van den Ban (2000b).

matching qualifications with the positions required, clear polices on recruitment, selection, placement and promotions, and in-service training.

Human resource development (HRD) has been much talked about and the time is ripe for a continuous HRD programme to be put in place. But, it is equally important to emphasize that, unless reward structures are right, and unless the trainers themselves have the prerequisite knowledge and skills, little improvement can be expected from a staff-training programme. In several cases, training posts are being filled through promotion from the lower ranks of DoAs. Where there is freedom to appoint from outside the organization, the low status and salary of the appointments has deterred potential applicants. Setting up autonomous training bodies, as suggested by DAC (2000), may be an interim step towards solving this problem.

## CONCLUSIONS

Despite an array of positive examples and forward looking policies, extension reform has yet to become widespread in India. The reasons for this are summarized in Box 2.8, and possible future roles and competence for extension are summarized in Figure 2.2. Box 2.8 provides an overview of how extension might respond to the particular challenges of weakly-integrated areas.

The major factor underpinning the above generic problems is the lack of agreement on the goal of extension, the objectives needed to achieve this goal, and the role that different organizations can play in the pursuit of these objectives. Many of the organizations that are already involved have too narrow a view of extension, and this limits their ability to mobilize around farmer needs on the one hand, and explore opportunities for collaboration and synergy on the other. Extension is still viewed as the transmission of technology, information, or materials such as seeds to farmers in order to introduce technical changes in agriculture that are considered desirable, without taking into consideration the livelihood options and resources of farmers. This inhibits any encouragement for farmers to reduce their dependence on public extension (van den Ban and Hawkins 1998). But the role of extension is much wider, since it also needs to teach management and decision-making skills, help rural people develop leadership and organizational skills enabling them to organize better, operate and/or participate in co-operative societies and other support

---

BOX 2.8
Constraints to Extension Reform in India

Constraints to reform include:

- the large size and diversity of the country; the division of responsibilities for research and extension between Centre and states in ways that impedes feedback on new requirements, responsiveness to these, and accountability to the users of new technology;

- deep-rooted perceptions of social status, that place research above extension, and many categories of the rural poor (SCs, STs, other backward castes or OBCs) at the bottom of the hierarchy, thereby limiting the effective interaction among them;

- the parallel operation of publicly funded, publicly implemented state extension services, all operating in isolation from each other, so that cross-learning is minimal, and the fact that they all engage permanent civil servants at all levels of the extension hierarchy;

- the lack of institutional learning, so that there are no consistent means of incorporating lessons learned by VEWs into new practice or policy;

- the prevalence of civil service behavioural norms across the hierarchy, including the pursuit at all levels of what may be locally inappropriate targets, the rigid interpretation of norms, leaving local workers little room for manoeuvre, the absence of substantive rewards linked to performance in responding to clients' needs, frequent transfers, and reluctance to serve in what are perceived to be punishment postings;

- the absence of social science skills among extension agents; and

- the absence of mechanisms to protect farmers from unscrupulous behaviour among private traders, and (as yet) of self-regulatory provisions among, for instance, input suppliers.

---

organizations, and to participate more fully in the development of local communities (Swanson and Clarr 1984).

The National Institute of Agricultural Extension Management has recently articulated that public sector extension in India, in addition to technology transfer, should embrace other roles such as HRD, broad-basing and farming system perspectives, and gender differentiated-strategies (MANAGE 1993). But, on-the-ground attempts to embrace any functions other than transfer of technology are few and dispersed and are likely to remain so in the absence of wide-scale upgrading of extensionists' skills, and for as long as administrative demands are high. Department of agriculture field workers are implementing programmes for the distribution of subsidized inputs[19] and have little time left for analytical field visits or participatory problem solving with farmers. This is not a surprising outcome since it has been estimated that administration for scheme implementation and meetings with higher officials consume 60 per cent of the working day for both agricultural officers and assistants (Jinraj 1999). The much-publicized group approach embraced by the DoA resulted in the formation of a number of such groups, but most of these remain dysfunctional and inactive due to a lack of clarity of purpose, and inadequate follow-up support for the activities of the groups (Jinraj 1999). This too is not surprising since staff do not have training in the sort of business and social science skills that makes support effective. Organizations that have been successful in this regard, for instance the KHDP, have specifically recruited personnel who could provide this expertise and have also contracted out training for group members (for more information see Isvarmurti 2000).

Though the availability and diffusion of appropriate technologies continue to be a challenge for the extension system, in the majority of rainfed and other disadvantaged areas, the wider adoption of technology necessitates collective action by farmers. Since many technologies suited to rainfed agriculture are knowledge-based and need community action, for example, integrated pest and common property resource management, farmer groups have to be sustained at the grassroots level. This is also essential if part of the extension function is to be transferred to farmer groups in the long term. Unless extension expands beyond the transfer of technology (ToT) mode, its relevance and utility for farmers will remain in doubt, and public support and commitment will decline further.

According to van Beek (1997), extension needs to move beyond ToT to embrace such roles as problem solving, education, and human development.[20] None of these four functions should be left out when designing or evaluating an extension project. Extension needs an increasing level of people oriented skills as it moves along the continuum from ToT to human development. Debate on the extent to which the Indian extension system is prepared to take on these responsibilities is long overdue.

India's acceptance of a World Bank/International Monetary Fund (IMF) Economic Reform Programme in the early 1990s, stimulated a wide-ranging reconsideration of the role of the state and of fiscal regimes (including subsidies). Much of this debate was limited to the industrial and 'modern' service sectors of the economy, and even here, there has in practice been only a limited rolling back of restrictions on private sector development. Saxena (2001) outlines some of the remaining restrictions affecting agriculture, and, certainly in relation to small-scale and marginal farmers, the old caricatures of near-helpless farmers, villainous private sector, and benevolent state remain firmly fixed in policy design and implementation. It will be many years before these begin to break down and a well-regulated private sector in areas such as seed provision comes to be recognized as being potentially of benefit to the rural poor.

Within public extension, ToT remains the dominant paradigm. Debates on knowledge system approaches in extension brought into focus the diverse sources of information and innovations and the need for stakeholders to be involved in all phases of the development of an innovation. This has led to the development of a number of participatory approaches as an alternative to the dominant ToT approach (discussed in detail by Chambers 1993). This triggered a large number of experiments in farmer participatory research and extension (FPR&E) or participatory technology development (PTD). Despite such experiments, agricultural scientists all too frequently find themselves struggling to apply participatory approaches in an institutional and professional context that pays only lip-service to such patterns of interaction with clients (Hall and Nahdy 1999). Though extension could potentially have contributed to such experimental approaches as PTD, the social distance between the research and extension systems has so far prevented them from undertaking any joint activities. Much is said about the importance of involving clients, but, in practice, such involvement is often a token gesture. Commonly, farmers are invited to meetings and are made members of some committees.[21] Peoples' representatives have been used to guide and monitor the activities of public sector extension in states such as Kerala. The ATMA also provides opportunities to bring client needs into the system through FIACs. They are expected to increase accountability but in societies that are deeply divided among communities and castes, this type of client representation is unlikely to be truly inclusive. In most regions and for most crops, clients and their representatives are in any case too weak to articulate their concerns. The public sector has a primary duty to give clients a voice through the formation of strong farmers' organizations but, in reality, the capacity of the public sector to help in organising farmer groups has proven weak, and there are not enough NGOs to take on this role in more than a small percentage of the areas that require it. The 'ideal' requirements for extension to serve weakly-integrated areas are well-known (Box 2.9); but there are problems in progressing towards these given the institutional rigidities in India.

---

### Box 2.9
### Extension in Weakly Integrated Areas

For areas weakly integrated into markets, the following would represent an ideal extension scenario:

- recognition by agricultural policy that, though productivity enhancement is important, it needs to be designed and implemented in ways that reduce vulnerability and create employment;

- recognition by rural development policy that agriculture alone (even in the broad definition used here) will be insufficient to generate the wide range of livelihood options that would be of value to the poor, so that public resources would be invested in the development of small and medium scale towns; and in transport and telecommunications infrastructure so that the poor may easily access employment opportunities and services (whether or not agriculture-related) offered by the government and private agencies in these towns;

- the strengthening of high-level skills in particular agriculture-related subject areas in small and medium towns, but with all specialists located in a single office and with close links to the private sector, so that visitors can easily be guided to the most appropriate source of information;

- close integration between such information sources and the providers of training, so that training curricula can constantly be updated to reflect the types of information that farmers request;

- efforts to build capacity among the poor to enhance their participation in the design of technology change programmes and to make demands on extension services, and input suppliers;

- outreach from small towns to villages to support farmer-to-farmer exchange of information and experience, and farmer experimentation;

- a switch in public funds out of the better-integrated areas (where there are good prospects for much extension to be privately supplied and funded) into weakly integrated areas where market failures predominate;

- the identification of market failures in such areas that the public sector has some prospect of rectifying (and over what period), plus the identification of what will remain public goods in both weakly and well-integrated areas (information and training on soil and water conservation and other environmental concerns, together with health, nutrition, and safety matters are likely to fall into this category);

- the development and implementation of a strategy to redress market failure over a defined time period and to provide public goods in ways that respond to the needs of the poor; and

- the replacement of civil service VEWs by private extension agents at the village level, who nonetheless will be publicly funded to some degree.

The need to address institutional dimensions of technology development and the importance of an inclusive analysis of knowledge flows and levels of interaction among all the actors in the innovation system[22] became more evident in the late 1980s. The emergence of the National Systems of Innovation (NSI) framework (Freeman 1987; Lundvall 1992) was a response to this issue. Analysis based on the NSI approach stresses that it is the performance of the system as a whole that is important, success tending to be a function of interaction and interactive relationships—often partnerships—that determine the effectiveness of knowledge flows between institutional nodes.

Several of the institutional innovations that have emerged in response to weaknesses in public sector research and extension, point to the emergence of an agricultural innovation system in India. This has resulted in blurring of the clearly demarcated institutional boundaries between research, extension, farmers, farmers' groups, NGOs, and private enterprises. Extension has a very important role to play in facilitating the nodes to generate, access, and transfer knowledge between different entities in the innovation system. It also has to create competent institutional nodes to improve the overall performance of the innovation system. Inability to play this very important role will further marginalize extension.

A consultation document recently released by the GoI/DoA (Box 2.5) contains provisions that correspond closely with the above, except that it sees little movement away from productivity-dominated agriculture policies. The main question that concerns us here is how easily implemented would such an approach be. Certainly, the evidence reviewed here suggests that progress towards an ideal of this kind is likely to be slow in India. In conclusion, we reiterate the reasons for this, indicating likely patterns of change (and resistance to change) and suggesting a number of second-best options that might be pursued at various points.

The post T&V period has seen experimentation with diverse extension approaches by a large number of extension providers. But most of them have not addressed the generic problems facing extension in India. The basic issue underpinning many of these has been the lack of a clear articulation of the role of extension in the Indian context. Public sector extension has to look beyond ToT roles. With the changing developmental agenda, extension in India will have to devise strategies for facilitating the poor to pursue broader livelihood options in on-farm and non-farm sectors so that their vulnerability is reduced. There is clear recognition in the Department of Agriculture and Co-operation consultation document of the need to adopt new visions for extension along these lines, to use Block towns (rather than VEWs) as the locus for improved services (that can be closely linked with private sector activity), to engage in multi-agency partnerships, and so on. However, this vision faces severe and longstanding implementation problems. Given the complexity and intractability of these, a wide-scale transformation of what is still predominantly publicly funded and publicly implemented extension in India is likely to take at least a decade. The poor would benefit substantially from efforts by innovative civil servants, NGOs, and donor-supported projects to accelerate this process of change.

## Endnotes

1. The average excess of actual procurement prices announced for wheat over the cost of production during the 1980s was 63 per cent, which increased to 96 per cent during the 1990s. A similar trend is observed in the case of rice.

2. Some estimates, for instance, indicate that only 30–40 per cent of water entering irrigation channels actually reaches end-users. One estimate is that a 10 per cent improvement in water management efficiency would add 14 million hectares to the total irrigated area (OECD 2000).

3. This includes the cost of subsidizing the producer price of food through the Guaranteed Purchase System.

4. The interventions were built on three foundations, namely improved package of farming technologies, a system of supply of critical modern inputs, and a remunerative price and market environment for farmers.

5. Rainfed agriculture covers 63 per cent of the total cultivated land and accounts for 45 per cent of agricultural production.

6. There is much debate over this figure, some estimates suggesting that 30 per cent of the population is still below the poverty line (Deaton and Drèze 2002).

7. According to Saxena (2001), this may be due to sluggish agricultural growth which is also poorly distributed spatially; inadequate reach of the Targeted PDS (TPDS) to the poorest in the northern and eastern states; the limitations of watershed development and poverty alleviation schemes; fiscal crisis caused by awards under the Fifth Pay Commission that led to reduced ability of states to spend on social sectors and on maintenance of assets; and deteriorating governance leading to leakages and inefficient utilization of resources.

8. According to a World Bank (2000) country study on India, the growth of real daily wages in rural areas—a key link between agricultural growth and poverty reduction according to the analyses of the 1980s—slowed down in the 1990s.

9. Workshop 18–19 January 2001 convened to elicit the views of State governments on the DAC policy document.

10. For a detailed discussion on investments and performance of various extension organizations in India see Sulaiman and Sadamate (2000).

11. Extension provision is constitutionally mandated to the states, not to central government. The great majority of extension agents are permanent civil servants, and attrition through natural turnover or redeployment is likely to be slow.

12. For instance, public sector extension workers in several states have been required to work through farmer groups in recent years, but groups are often formed on a token basis, simply so that a list of names can be produced by the VEW when required by his supervisor.

13. Public goods are those having low subtractability and excludability, whereas private goods are those having high substractability and excludability. Excludability applies when access is denied to those who have not paid for the product, while substractability (rivalry) applies when an individual's use or consumption of a good or service reduces its availability to others. Purely public and private goods occupy opposite ends of the economic spectrum. Between the two extremes, are toll goods and common-pool goods.

14. By contrast, managed forms of triage would, for instance, monitor how adequately such labour is being absorbed elsewhere in the economy and would seek to provide incentives or restrictions such that the volume of labour flows is consistent with available opportunities.

15. 18 districts having assured water supply, minimum natural hazards, well-developed institutions such as co-operatives and panchayats and maximum potential for increasing agricultural production within a short time were selected for implementing the programme.

16. One of the reasons for the mounting food stocks with the government has been the low demand for these grains, as the poor (and poor states) lack the purchasing power to buy them. It has been estimated that to deliver Re 1 of

subsidized food to the poor costs an additional Rs 1-2, depending on the extent of 'leakage' (World Bank 2002).

17. Swanson (1996) consider that a fully functional extension system needs to have 30-5 per cent of its total expenses as operational expenditure.

18. DoA staff need skills related to group formation, leadership development, conflict resolution, inter-group negotiation, and management of common property resources. For more details see Farrington *et al.* (1998).

19. Most of the central and state sector agricultural development programmes have a component for providing subsidized inputs or subsidies to participating farmers. For instance, seeds/seedlings (promotion of new varieties or crops), pesticides and equipments such as sprayers (control of some crop specific pests), bio-fertilizers (promotion of bio-fertilizer application) etc. are provided through the DoA. Cash subsidies are also provided by DoA for construction of wells, implementation of drip irrigation, and purchase of equipment such as tractors.

20. van Beek (1997) has defined these four roles as follows: technology transfer which links research with users; problem solving, which assists clients with solving individual problems; education, which aims to empower people to solve their problems; and human development which encourages people to govern themselves and to develop their learning capability.

21. Usually these are well educated, resource-rich farmers who may not fully understand the problems of the resource-poor subsistence farmers.

22. An innovation system encompasses all the elements of the system or network of private and public sector institutions whose interactions produce, diffuse, and use economically useful knowledge. Therefore, this type of analysis is more inclusive than the narrower notion of a research or extension system.

## References

Alsop, R., E. Gilbert, J. Farrington, and R. Khandelwal (1999), *Coalitions of Interest: Partnerships for Process of Agricultural Change*, New Delhi and London: Sage.

Ameur, C. (1994), 'Agricultural Extension: A step beyond the next step', World Bank Technical Paper 247, Washington DC: World Bank.

Biggs, S. P. (1990), 'A Multiple Source of Innovation Model of Agricultural Research and Technology Promotion', *World Development*, Vol. 18, No. 11, pp. 1481-99.

Chambers, R. (1993), *Challenging the Professions*, London: Intermediate Technology Publications.

Chandra Shekhara, P. (ed.) (2001), *Private Extension in India: Myths, Realities, Apprehensions and Approaches*, Hyderabad: National Institute of Agricultural Extension Management (MANAGE).

Christoplos, I., J. Farrington, and A. D. Kidd (2000), *Extension, Poverty and Vulnerability—Inception Report*, Working Paper 144, London: Overseas Development Institute (mimeo).

Coutts, J. A. (1995), 'Agricultural Extension Policy as a Framework for Change', *European Journal of Agricultural Education and Extension*, Vol. 2, No. 1, pp. 17–26.

DAC (2000), *Policy Framework for Agricultural Extension*, Extension Division, Department of Agriculture and Co-operation (DAC), Ministry of Agriculture, New Delhi: Government of India (draft).

Das, P. and B. S. Hansra (1999), *Status report on Krishi Vigyan Kendras—A reality*, New Delhi: Indian Council of Agricultural Research.

Deaton, A. and J. Drèze (2002), 'Poverty and Inequality in India: a Re-examination', *Economic and Political Weekly*, 7 September 2002.

Farrington, J. (2001), 'Sustainable Livelihoods, Rights, and the New Architecture of Aid', *Natural Resource Perspectives* 69, London: Overseas Development Institute (ODI).

Farrington, J., I. Christoplos, A. Kidd, and M. Beckman (2002), 'Poverty, Extension and Vulnerability: The scope for policy reform', Final report of a study conducted on behalf of the Neuchâtel Initiative, ODI Working Paper 155, London: Overseas Development Institute.

Farrington, J., N. C. Saxena, T. Barton, and R. Nayak (2003), 'Post Offices, Pensions and Computers: New Opportunities for Combining Growth and Social Protection in Weakly Integrated Areas', *Natural Resource Perspectives No. 84*, London: Overseas Development Institute.

Farrington, J., V. R. Sulaiman, and Suresh Pal (1998), 'Improving the effectiveness of agricultural research and extension in India: An Analysis of Institutional and Socio-economic Issues in Rainfed Areas', Policy Paper 8, New Delhi: National Centre for Agricultural Economics and Policy Research.

Freeman, C. C. (1987), *Technology and Economic Performance—Lessons from Japan*, London: Pinter.

Gill, D. S. (1991), 'Economic returns to expenditures on agricultural extension system', *Journal of Extension Systems*, Vol. 7, pp. 44–61.

Hall, A. J., N. G. Clark, S. Taylor, and R. V. Sulaiman, (2001), 'Institutional learning through technical projects: Horticulture technology R&D system in India', ODI Agricultural Research and Extension Network Paper 111, London: Overseas Development Institute.

Hall, A. J., R. V. Sulaiman, N. G. Clark, M. V. K. Sivamohan, and B. Yoganand (2000), 'Public and private sector partnerships in Indian agricultural research: Emerging challenges to creating an agricultural innovation system', presented at the XXIV International Conference of Agricultural Economists, 13–18 August, Berlin, Germany.

Hall, A. J. and S. Nahdy (1999), 'New methods and old institutions: Systems problems of farmers participatory research', ODI Agricultural Research and Extension Network Paper 93, London: Overseas Development Institute.

Haque, T. (2000), 'Status and strategy of agricultural development in backward districts of India', New Delhi: National Centre for Agricultural Economics and Policy Research (mimeo).

ICAR (2001), *Annual Report 2000–2001*, New Delhi: Indian Council of Agricultural Research.

———— (1998), *National Agricultural Technology Project: Main document*, New Delhi: Indian Council of Agricultural Research.

———— (1996), 'Report of the review committee on extension systems of the ICAR (Surjewala Committee Report)', Indian Council of Agricultural Research, New Delhi: Division of Agricultural Extension.

Isvarmurti, K. (2000), 'Kerala Horticulture Development Programme—New way to help farmers', *Agriculture and Industry Survey*, Vol. 10, No. 11, pp. 23–34.

Jha, D. and A. Kandaswamy (eds) (1994), *Decentralising agricultural research and technology transfer in India*, New Delhi: Indian Council of Agricultural Research (ICAR) and Washington, DC: International Food Policy Research Institute (IFPRI).

Jinraj, P. V. (1999), 'Performance Evaluation of Krishi Bhavans set up in Kerala', Discussion Paper 8, Thiruvanthapuram: Kerala Research Programme on Local Level Development, Centre for Development Studies.

Kaimovitz, D. (1991), 'The evolution of links between research and extension in developing countries' in W. M. Rivera and D. J. Gustafson (eds) (1991), *Agricultural extension Worldwide institutional evolution and forces for change*, Amsterdam: Elsevier.

Lundvall, B. A. (ed.) (1992), *National Innovation System: Towards a theory of innovation and interactive learnings*, London: Pinter.

Macklin, M. (1992), 'Agricultural Extension in India', World Bank Technical Paper 190, Washington DC: World Bank.

MANAGE (1999a), 'Innovations in Technology Dissemination', *MANAGE-NATP series 1*, Hyderabad: National Institute of Agricultural Extension Management (MANAGE).

———— (1999b), 'Agricultural Technology Management Agency'. *MANAGE-NATP series 2*, Hyderabad: National Institute of Agricultural Extension Management.

———— (1993), *Farmers Participation in Agricultural Research and Extension Systems*, Hyderabad: National Institute of Agricultural Extension Management (MANAGE).

Moris, J. (1991), *Extension Alternatives in Tropical Africa*, London: Overseas Development Institute.

OECD (2000), *Agricultural Policies in Emerging and Transition Economies*, Paris: Directorate for Food, Agriculture and Fisheries, Organization for Economic Cooperation and Development.

Rivera, W. M and D. J. Gustafson (eds) (1991), *Agricultural extension—*

*Worldwide institutional evolution and forces for change*, Amsterdam: Elsevier.

Saxena, N. C. (2001), 'How have the poor done? Mid-term Review of India's Ninth Five-Year Plan', *Natural Resource Perspectives* 66, London: Overseas Development Institute (ODI).

Sharada, H. C., P. Ballabh, M. Chauhan, and P. K. Labh (1996), 'Process Documentation of Client driven Agriculture research and extension— Experiences of VB KVK, Udaipur' in D. V. Rangnekar and J. Farrington (eds), *Process Documentation and Monitoring in Action*, Pune, India: Bharatiya Agro-Industries Federation (BAIF).

Sharma, R. (1999), 'Reforms in Indian Agriculture: The case of Agricultural Extension', Paper presented at the NCAER–IEG–WB Workshop on Agricultural Policies, 15–16 April 1990, New Delhi.

Sharma, V. P. (2000), 'Cyber Extension in the context of Agricultural Extension in India', *MANAGE Extension Research Review*, Hyderabad: National Institute of Agricultural Extension Management, pp. 24–41.

Sood, A. D. (2001), 'How to Wire Rural India: Problems and Possibilities of Digital Development', *Economic and Political Weekly*, Vol. 36, No. 43, pp. 4134–41.

Sulaiman, V. R. and V. V. Sadamate (2000), 'Privatising Agricultural Extension in India', Policy Paper 10, New Delhi: National Centre for Agricultural Economics and Policy Research.

Sulaiman, V. R. and D. Jha (2000), 'Determinants of demand for paid farm extension services in India: A discriminant functions approach', *Afro-Asian Journal of Rural Development*, Vol. 33, No. 2, pp. 57–67.

Sulaiman, V. R. and A. W. van den Ban (2000a), 'Reorienting Agricultural Extension Curricula in India', *Journal of Agricultural Education and Extension*, Vol. 7, No. 2, pp. 69–78.

————— (2000b), 'Agricultural Extension in India—The Next Step', *NCAP Policy Brief*, New Delhi: National Centre for Agricultural Economics and Policy Research.

Swanson, B. E. (1996), 'Innovations in Technology Dissemination component of NATP', Prepared for Ministry of Agriculture, Government of India, New Delhi. (mimeo).

Swanson, B. E. and J. B. Clarr (1984), 'The History and Development of Agricultural Extension', in B. E. Swanson (ed.), *Agricultural Extension— A Relevance Manual*, Rome: Food and Agricultural Organization of the United Nations (FAO), pp. 1–19.

van den Ban, A. W. (2000), 'Different Ways of Financing Agricultural Extension', *ODI Agricultural Research and Extension Network Paper* 106b, London: Overseas Development Institute.

————— (1998), 'Supporting farmers' Decision-making Process by Agricultural Extension', *Journal of Extension Systems*, Vol. 14, pp. 55–64.

van den Ban, A. W. and B. S. Hawkins (1998), *Agricultural Extension*, New Delhi: CBS Publishers and Distributors.
———— (1996), *Agricultural Extension, Agricultural Extension*, Oxford: Blackwell Science Publishers.
van Beek, P. G. H. (1997), 'Beyond Technology Transfer', *European Journal of Agricultural Education and Extension*, Vol. 4, No. 3, pp. 183–92.
World Bank (2002), 2002 World Development Indicators, CD Rom, Washington DC: World Bank.
———— (2000), *India—Reducing Poverty, Accelerating Development, A World Bank Country Study*, New Delhi: Oxford University Press.

# 3

# Extension, Poverty, and Vulnerability in Vietnam

*Malin Beckman*

## THE COUNTRY CONTEXT

### BASIC INDICATORS

Of a total population of over 78 million, some 70 per cent is rural. Landholdings, at 0.07ha per person, are exceptionally small, but irrigation is widespread and productivity is high in many areas. Despite some recent improvements, other social and economic indicators are generally poor, and are summarized in Table 3.1.

### RURAL POVERTY

The majority of the poor (90 per cent) live in rural areas, and poverty is concentrated in the northern hill areas (Map 3.1), mainly among ethnic minorities. In Vietnam, the nature of poverty is changing. Ten years ago, almost the entire population was poor, and policies that stimulated overall growth in the economy would almost automatically benefit the poor. Now there is a need for more carefully targeted measures of poverty alleviation. Poverty is increasingly connected with vulnerability. Vulnerable groups include ethnic minorities in the remote mountain areas, people in the disaster-prone areas of the north central coast, laid-off workers from the state-owned enterprises, single-headed households, victims of the war, and landless workers in the south.

In 1998, nearly 80 per cent of the rural poor in Vietnam worked on their own land. Of the 20 per cent without land, the majority are in the south. In north and central Vietnam, the rural poor usually

TABLE 3.1
Vietnam, Basic Indicators

| Series | Value | Year |
| --- | --- | --- |
| Cereal yield (kg per hectare) | 4048.5 | 2000 |
| Land use, arable land (hectares per person) | 0.07 | 1999 |
| Land use, irrigated land (% of cropland) | 41 | 1999 |
| Agriculture, value added (% of GDP) | 24 | 2000 |
| GNI per capita, Atlas method (current US$) | 390 | 2000 |
| Population, total | 78,522,704 | 2000 |
| Rural population (% of total population) | 76 | 2000 |
| Malnutrition prevalence, height for age (% of children under 5) | 39 | 1999 |
| Malnutrition prevalence, weight for age (% of children under 5) | 38 | 1999 |
| Low birthweight babies (% of births) | 11 | 1994 |
| Poverty headcount, national (% of population) | 51 | 1993 |
| Poverty headcount, rural (% of population) | 57 | 1993 |
| GINI index | 36 | 1998 |
| Mortality rate, infant (per 1000 live births) | 27.5 | 2000 |
| School enrolment, primary (% net) | 97 | 1998 |
| Surface area (sq. km) | 331,690 | 2000 |
| Roads, total network (km) | 93,300 | 1999 |

*Source*: World Bank (2002).

still have access to land and basic productive resources. In the south, there are many poor people who are landless labourers because of the skewed access to land. This is a historical difference. The land reforms in the north started early, following the creation of the Democratic Republic of Vietnam in 1945. There were land reforms in the south also, after 1975, but it has not been possible to enforce them as stringently as in the north. In recent years, some large landowners have managed to reclaim their land at the expense of the rural poor. This study is based on research conducted on rural poverty in north and central Vietnam, where the main household income still comes from farming, and where there is not yet any significant land concentration in larger farms. Households, which are expanding economically, do so mainly in terms of diversification into business development, rather than in expanding landholdings.

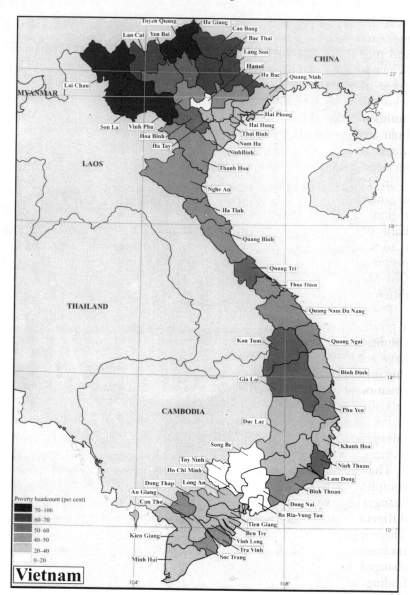

1994 MAGELLAN GeographixSM Santa Barbara, CA (800) 929–4627

MAP 3.1: Spatial distribution of poverty in Vietnam by province

*Source:* Based on Minot, N. and B. Baulch (2000), 'The spatial distribution of poverty in Vietnam and the potential for targeting', World Bank/IFPRI.

Although the rural poor to a large extent are own-account farmers, there are also other sources of income, including labour opportunities and migration, which play a vital role in their livelihood security. When looking at poverty from a livelihood perspective, there is a range of non-production factors, which are crucial in determining who will become mired in poverty or destitution. Such factors include health problems, causing large costs and labour shortages for household activities, and relatively lower education. This tends to reduce access to markets and services for production, including advisory services.

In 1999 the World Bank, in co-operation with other donors organized participatory poverty assessments (PPAs) in four different regions. These are compiled in the report 'Voices of the Poor' (World Bank 1999a). The four cases include the northern mountainous region (Lao Cai province), the north central coastal region (Ha Tinh province), the southern lowland region (Tra Vinh province) and Ho Chi Minh City. The nature of poverty in the four areas is quite different.[1]

Poverty in the mountain areas is relatively homogeneous within a given community, and is largely related to the overall difficult conditions in which people live. In many areas, mountain dwellers have gone through major changes in their production systems, from shifting cultivation to fixed cultivation. This puts them in a vulnerable situation as their traditional knowledge with regard to cultivation, the spreading of risk, and adaptation to the environment is no longer 'valid'. The transformation to fixed cultivation is constrained by poor infrastructure and remoteness from district and provincial centres, services, and markets.

The situation in the lower mountain areas is often socially complex, with the immigration of people from the lowlands causing changes in land rights and practices. The immigrants bring with them different cultivation practices, at times causing conflicts over land. There is often insufficient knowledge among policy makers about the traditional systems of access to land. The traditional systems have also changed with the move from shifting cultivation. The process of finding appropriate forms for land allocation in the mountains is still going on.

Poverty in the north central coastal areas is linked to difficult natural conditions—frequent floods and drought. Vulnerability to seasonal crises is the main cause of poverty. Poor households, which are solely dependent on the production of rice, with little diversification of income sources are especially vulnerable.

Poverty in the lowland delta areas is less homogeneous than that in the other areas. Poverty can be seen in the midst of better-off rural areas. Several problems can make it very difficult for a family to climb out of poverty. The young generation often have less access to land, inequalities with respect to access to irrigation and drainage, small amount of labour in relation to dependants in the family, negative debt spirals, health problems, and relatively lower education. On the other hand, these lowland areas are well integrated in the market and the majority of the population produces a surplus. Poverty in these areas is, therefore, often given less priority in local, national, and international development agendas, which tend to focus on the remote and non-market integrated areas.

## POLICIES FOR AGRICULTURE AND RURAL DEVELOPMENT

### Poverty Reduction

Poverty reduction is the overarching policy goal in Vietnam. Poverty in Vietnam has been substantially reduced during the past decade, from 58 per cent in 1993 to 37 per cent in 1998, according to World Bank statistics. Agricultural incomes rose by 60 per cent during the same period. Living standards are rising, which is also reflected in other indicators of human development. School enrolment rates have improved, as well as indicators such as access to household assets, for example, television sets and bicycles (World Bank 1999b).[2]

Food security has been the main poverty reduction objective during the 1990s. Starting from a situation of severe lack of food in the first decade after the American war, the scene has now changed to relative food security for the majority of the population as well as significant export production. Rice production is the basis of food security in most rural areas, but its decreasing profitability has led to the recognition that further increases in rice production should no longer be the main focus for poverty alleviation. The policy emphasis for 2000–10 lies on market orientation and raising living standards for vulnerable groups especially in the remote minority areas.

Growth has so far been driven by the rural sector. The primary route to overall poverty reduction is perceived as being broad-based economic growth plus structural reforms intended to promote employment and exports. The relatively even distribution of resources provides the basis for a positive relation between general economic growth and poverty reduction. There is, however, an increasing recognition of the need for measures targeted more at vulnerable groups, including ethnic minorities, female-headed households, migrants to urban areas, and people in disaster-prone areas.

The growth and poverty objectives are brought together in a 'Comprehensive Poverty Reduction and Growth Strategy (CPRGS) prepared by the Government of Vietnam in co-operation with the World Bank and other international organizations.

Three broad imperatives have been set out in the fight against poverty:[3]

1. creating opportunities for employment and productivity growth, in order to raise incomes for the poor;

2. adopting measures to ensure that growth and access to services are fair and equitable; and

3. reducing the vulnerability of the poor to events such as sickness and crop losses.

The CPRGS takes a very market-oriented approach to poverty alleviation and comprises broad outlines for the liberalization of the economy to achieve rural economic growth. This includes providing equal access to capital, credit, land, labour, technology, information, as well as government incentives for all types of enterprises, liberalizing the trade and banking system. The focus is on policies to create a positive business environment and to support the development of small and medium-sized enterprises.

In parallel, the CPRGS proposes increased government investment in sectors considered important for poverty alleviation, including agricultural extension services, water management, agro-processing, education, training, and health. It is suggested that the poor should receive preferential treatment in terms of subsidized credit, extension services, and the like. In the long run, however, the formal credit system should be adapted to become more accessible to the poor by means of appropriate procedures and loan conditions rather than subsidies, though special credit conditions are required to help the poor cope with market difficulties, such as falling prices. Research on appropriate technologies for the poor is to be supported. Market centres are to be developed in remote areas to facilitate the exchange of information and products.

The Bank for the Poor was established in 1994 and has distributed credit at a subsidized interest rate to a high proportion of the poor households in the country. People in remote communes have had less access to such credit, however, and the credit has mostly been allocated without the back-up of advisory services.

From the poverty perspective, there are tensions between liberalizing markets and the immediate situation of the poor. Fertilizer subsidies are being removed, which raises the costs of rice production; at times, input costs are even higher than the value of marketed rice. As subsidies to the state-owned enterprises are dismantled, large numbers of workers become unemployed. There is a general emphasis on the need to steer away from crops which have no comparative advantage. Market forces are, however, not being given completely free reign. There is a continued commitment to support the troubled sugar industry, for example, in order to protect current investments and employment.

Development in the mountain areas can be seen from two perspectives. On the one hand, these are the poorest areas in the country and massive efforts are now being made in order to raise living standards in the mountains in line with those of the rest of the country. A comprehensive state programme[4] (partly with international support) is investing in infrastructure for both production and social services. On the other hand, the mountain areas are seen as highly interesting from an export market perspective, despite their remoteness. These two perspectives can at times be in conflict. Provincial governments often embark upon large-scale export production strategies without sufficient development of market channels. These high-risk projects frequently weaken local food security strategies. The CPRGS mentions support for specialized commodity production areas, based on each region's particular endowments but insufficient attention is paid to developing sustainable local production systems for food security to reduce vulnerability to changes in export market conditions.

RURAL LABOUR MARKETS

Employment creation policies in the rural areas mainly target the large rural workforce who are underemployed on their small landholdings and who have a large demand for supplementary income-generation activities and small business development. The emphasis is on supporting rural industrialization based on agro-processing and small-scale rural enterprise, both on and off-farm.

The main objective for the rural population as well as policy makers is to create conditions for the farmers to continue to have land based production as their base (not necessarily their main income), and to supplement it with small business development, employment, and other off-farm income. Poor farmers take on seasonal employment to supplement their own-account farming. Also full-time employment is regarded as supplementary income as sons and daughters seek employment outside the farm, in order to support the family.

Most technologies and extension messages in agriculture are adapted to the own-account producer and are designed to be labour-saving, as labour is a scarce resource in family agriculture. (Rice production is extremely labour intensive.) The technique of throwing rice seedlings instead of planting them, which is widespread in China, is now also spreading in Vietnam. It is highly labour-saving at the planting stage. Tractor ploughing is gradually taking over from buffalo ploughing. From an employment perspective, this reduces opportunities during the peak season. The main demand for employment is, however, still during the off-peak season, as the majority of the rural labour force also have their own plots to tend to.

Government rhetoric favours the development of large-scale commercialized farms (*trang trai*). Concrete proposals in this direction have, however, been turned down by the National Assembly, as the vast majority in north and central Vietnam are still smallholder farmers. A few large-scale farms, with capital-intensive production, are emerging in peri-urban areas, supplying the urban markets with meat and vegetables. The government programme for rural employment creation, ('Decree 120') involves the provision of credit funds and vocational training to encourage small-scale rural enterprise. So far, a large proportion of the funds go to animal husbandry, which has been the most common way for farmers to diversify their income. There is an ongoing local political struggle over whether to favour large-scale animal husbandry as employment generation or to spread out support to many small-scale producers. The government has introduced subsidies for households that are prepared to invest in farms with 50 pigs or more. There is disagreement about whether this should be funded through the poverty alleviation programmes, as large-scale pig production may compete with small-scale producers. The government claims that this is not a problem, as large-scale production is aimed at urban rather than local rural markets. This argument may hold in the long run, but at present the large-scale producers are dependent on local markets while 'waiting for' veterinary standards to be improved and a certification system to be established as these are required for expansion to urban markets.

Non-agricultural, small-scale enterprises are becoming more common, but there is a lack of advisory services to provide the knowledge and information that people require in order to venture into new types of enterprise.

Migrant labour is an important supplementary income opportunity for the poor. In the disaster-prone areas migrant labour is also a coping strategy for a large part of the population after a crisis. Seasonal migration of labour is common in areas producing export crops. Many people used to migrate seasonally to coffee-growing areas in the central highlands, but with coffee prices having slumped in recent years, that option is no longer available. Industrialization is stronger in the south and many young people from the rest of the country migrate to the south to seek employment in manufacturing industries, mainly the textile industry in Ho Chi Minh city. Migration is mostly seen as a supplementary income opportunity rather than as an alternative livelihood strategy. There is, however, also a policy of encouraging resettlement of whole communities when an area becomes overpopulated or is otherwise unable to support its population. Such resettlement has mainly been from the lowlands to the mountains.

## VULNERABILITY AND DISASTER MITIGATION

Although achievements in poverty reduction have been considerable, there are still a large number of people who are vulnerable to crises, which could push them back into poverty. Such crises may be crop losses due to floods and drought, illness in the family—giving rise to both medical costs and reduced labour capacity, loss of buffalo or other livestock, or falling market prices.

Floods and drought occur frequently in many parts of the country and contribute to a general level of vulnerability. At times they are more severe than normal and turn into disasters, such as the floods in central Vietnam in November 1999, when 2700 mm of rain fell in four days. Over 700 people lost their lives and there were huge material losses. In the summer of 2000, the Mekong river overflowed and flooded the lowland of the Mekong delta for three months, causing high human and material losses. Reducing vulnerability has thus been placed high on the political agenda in the last few years.

---

Box 3.1

Policies to Reduce Vulnerability

• water management infrastructure to reduce the impact of floods and drought;

• diversification of production to reduce the impact on the household economy if one line of production fails;

• community savings and credit schemes to reduce dependence on private money lenders;

• insurance services for production losses;

• improved health insurance systems;

• improved veterinary services to reduce the occurrence of diseases in animal husbandry due to floods.

---

The immediate government support has concentrated on safeguarding food security and production in the short run and strengthening infrastructure for disaster mitigation. Extension services have increased their focus on developing short-term varieties of rice and other crops, which are thus less exposed to the flood risk. The vulnerability which comes from mono-cropping and dependency on one major source of income is increasingly being recognized. It remains to be seen whether this will be taken into consideration in production planning, as the government still has a strong role in encouraging certain types of production, and gearing resources and subsidies towards chosen production priorities. So far these production recommendations have favoured increasing bulk production in order to access new markets, thereby encouraging farmers to concentrate resources on that particular product, which in itself increases vulnerability to both weather problems and market changes.

Reducing vulnerability in the mountain areas involves a certain contradiction caused by the need to reduce vulnerability for the lowland population by policies for overall forest cover. The mountain population, however, is dependent on access to the mountain slopes for cultivation for food security. The 1999 floods were a disaster for the mountain population in Thua Thien Hue province as they had followed policy directives and moved all their production and residential areas to the riverbanks, away from the slopes. The production on the riverbanks was heavily affected by the force of the river, which eroded land and deposited stone and sand on the fields.

The compromise solution that is being argued for by the mountain population and local government includes various forms of agro-forestry production on the slopes. So far there has been insufficient policy and advisory back-up for such solutions.

Findings from a study on. recovery after the floods in 1999 (Beckman *et al.* 2001)—see also Boxes 3.1 and 3.2—suggest that the people with most diversified household economy had the best potential to recover quickly after the floods. Even though villages in the lower hills are generally poorer than those in the lowland paddy areas, the hill dwellers had better conditions for recovery due to access to a range of small income opportunities from minor forest products

---

BOX 3.2

Floods and Vulnerability

We are five in the family, and my wife is the only main worker. I have been handicapped since the American war and receive a state pension of 360,000 dong per month. A lot of that money goes for hospital care of our child who fell into a pond.

We have one sao (500 sq.m) of maize and beans for the winter crop to provide for ourselves and fodder for the chickens. We lost 30 chickens in the floods. After the floods we had 16. Now 10 chickens remain. There are a lot of diseases and we cannot afford vaccination. In 1998–9 we raised sugarcane. We did not get any payment for the first harvest because the man who received the sugarcane lost the receipt. The second harvest was largely damaged by flood, but it did not make any difference because no transport came to collect the harvest anyway. After that we changed to cassava instead. We planted 2000 cassava plants, but they were inundated by the heavy rains. The roots went black and rotted. We have planted some dry land rice (local seed), which we have not yet harvested. We hope this harvest will be good. We borrow land from relatives. We have bananas and vegetables around the house. We planted forest in contract with government programmes, between 1994 and 2000, a total of 4.8 ha. We were paid for planting and tending for three years but most of it is finished now. We would like to plant fruit trees, but do not know how to get seedlings. We received credit to buy a sow, but it died. We do not know how we will pay back this debt and the debt to the government for sugarcane inputs (Quynh Gia Eo and Ho Thi Hat, Hong Ha Commune, A Luoi District, Thua Thien Hue Province).

*Source:* Beckman 2000 (fieldwork notes).

to alternative labour opportunities. In this case, however, the hill area happened to be closer to highways and markets than the lowland area, which played a significant role for diversification opportunities.

The study concludes that the possibilities of recovering from a disaster are determined more by the general household livelihood situation, than by the size of the loss itself. The concept of access is crucial for resilience. In remote areas, the lack of access influences both the overall livelihood development opportunities and the access to resources to recover. Also in the 'well-integrated' areas, there is a need to be sensitive to poor people's lack of access to markets and services, which may be physically close, but socially still remote.

DECENTRALIZATION AND GOVERNANCE

In many countries decentralization is perceived to be constrained by a lack of capacity at the local government level. In Vietnam, the capacity of local government is often greater than the level of responsibility that is formally given to it. The demand for decentralization comes largely from below. At the district level, people's committees have the primary responsibility for local development, with accountability to both the people of the district and to the provincial and national governments. People in most districts have real possibilities of exercising pressure and influence at the district level through the commune people's committee and the people's council structures. Direct accountability to the people is, however, weaker higher up the hierarchy. Policy from national and provincial levels is not always appropriate or realistic at the local level, which puts the local government in difficult situations of balancing pressures from above and below. In general, however, there is a common policy focus at all levels, oriented towards rural development. In the district allocation of resources, the rural communes have strong power of negotiation, as they clearly outnumber the urban centres.

At the commune level, there are two main structures through which people can articulate their demands for public services. First, through the mass organizations, mainly the women's union and the farmers' association. They organize regular meetings in which everyone can take part and where ongoing development issues can be discussed. The second is the structure of hamlet, village, and commune decision making. Previously, these government-led structures were used mainly for spreading directives, but now they are increasingly becoming forums for development planning. One constraint for the poor in the lowland and midland areas is that are often in a minority within the commune, while the development decisions tend to focus on issues of priority to the majority.

The links between government and the communities tend to be weaker in the mountain areas, partly because of the physical distances, and also because of the traditionally different organization of authority in areas with ethnic minorities. The district people's committee also tends to have less overall responsibility and control due to the more diverse spectra of local, provincial, national, and international interests including state-owned enterprises and management boards for natural forests.

There is an ongoing process of public administration reform aimed at strengthening the supervisory and monitoring responsibilities of local government at the commune level. Funds for Programme 135 for the remote mountain communes are allocated on the basis of commune development plans, which increases the potential for cross-sector co-ordination and a strong role for the commune in the monitoring and implementation of activities and investments. The grassroots democracy decree (number 29, 1998) also contributes to this development by emphasizing people's right to information and to comment on and contribute to development plans and projects.

## AGRICULTURAL EXTENSION: BACKGROUND AND STATUS

### PLURALISM UNDER GOVERNMENT CO-ORDINATION

At the district level, Vietnam follows a collaborative model of extension, where local government has overall control over extension activities in the district and funding is primarily from public funds. The range of actors in extension is broad and includes public extension and related services, mass organizations, village organizations, farmer groups, co-operatives, private entrepreneurs, state and private input supply companies etc.

Many of the actors in this pluralistic model are community organizations with strong links to the state. The distinction between state and civil society in Vietnam is blurred. The various organizations operate with considerable independence but are still accountable to the local people's committee. There are advantages and disadvantages with a close relation between community organizations and the local government. It means that there is an institutional structure for communication between supply and demand for services. People have a better chance of influencing supply by voicing demand within the mass organizations than from outside. There is, however, the risk that the mass organizations spend most of their time mobilizing in support of state policy decisions already made, at the expense of their role as a channel for popular demand.

Private sector extension is growing, as policy moves towards creating more equal terms of competition between public and private enterprise. Subsidies to state companies are gradually being removed. Private extension comes mainly from small local providers of veterinary and plant protection services operating with state certification. The larger input supply companies often 'contract-in' public extension services for demonstrations and marketing of their products. The public services function as a quality guarantee between the private service providers and their customers. Ministry policy is to gradually move towards cost-sharing of extension services for commercial production and free services in the remote mountain areas. The implications of this policy for potential impact on the poor in commercial areas and for non-commercial extension are as yet unclear.

The Vietnamese concept of 'socializing' extension (*xa hoi hoa*) means that extension is in fact a responsibility of society as a whole and that all organizations play a part in raising production, improving technologies, and spreading production knowledge. In practice, this policy is mainly implemented through mass organizations, which also have important extension functions. Part of the national budget for extension is also allocated to the mass organizations for joint programmes with the line-ministry extension service.

The main line-ministry extension service under the Ministry of Agriculture and Rural Development (MARD) is in this chapter referred to as the 'Extension Organization' (*Khuyen Nong*). It is, however, only one of the organizations involved in extension activities. Under the MARD structure are also the Plant Protection Organization, the Veterinary Organization, the Forestry Organization, and the Fisheries Department, all of which undertake extension activities. Several other ministries are involved through development programmes, which also include extension components.[5]

At the district level, all line departments come under the district people's committee, which thus has a strong co-ordinating role; however, there is less co-ordination at the provincial and national levels. The co-ordinating mandate of the districts varies between provinces as each organization is also accountable to its respective departments at the provincial and national levels. The MARD guidelines favour strengthening the accountability of various organizations to the district people's committee, but in practice most of the strategic decisions are made at the provincial level, and some provinces are reluctant to devolve power.

## PRIVATE SERVICE PROVIDERS

Private service providers are to a large extent individual rural people, often farmers themselves, who provide veterinary services and plant protection services to the local population. They operate either independently or under contract to the co-operative or commune authorities. The local government agencies issue certification and operate quality controls. The level of knowledge is, however, uneven. More training and back-up is required for the private suppliers to function adequately.

The main providers of seed, seedlings, piglets, fingerlings etc. (and the advice required for their use) are from the private sector, except for the newest varieties and breeds, which most often come from government research stations. A few co-operatives have the capacity to produce rice seedlings for local supply. There are often pressures from the villages for the public sector extension services to support village level capacity to produce seedlings, in order to be more independent locally.

Private and public companies supplying inputs such as fertilizer and chemicals undertake extension activities to encourage the use of their products as well as demonstrating their appropriate use. Such extension is often in the form of demonstration models and seminars where farmers are invited to study various products and solutions. The input suppliers are keen to co-ordinate with the extension organization in order to give a more unbiased appearance.

Government extension staff often draw on their own professional knowledge and networks to provide private input services. This type of business may function as any other private supply, but at times there is an element of corruption. Government staff can contract each other for supply within government programmes. Kickbacks provided to the extension staff from such activities are less of a problem when dealing directly with the farmers, that is, the farmers use their own resources to buy the inputs. It is a more significant problem when the inputs are subsidized by government or donor funds.

It is becoming more common for groups of farmers to form interest groups or even economic organizations for the joint purchase of inputs, marketing, and the purchase of advice and technical support from government extension staff. It is normally the better-off farmers—with the capacity to take risks with new lines of production—who organize themselves in this way. The co-operatives and other village-wide organizations play an important role in broadening economic co-operation to also include poor farmers. Co-operatives may purchase inputs on behalf of the whole community and thereby negotiate better price and credit agreements. They may also be input providers themselves, making the supply business commercial, and owned by the members.

## VILLAGE-LEVEL EXTENSION

There is increasing recognition of the need for village extension networks to reach the poor. It is, however, unrealistic to rely on formal extension staff to provide direct advisory services to the poor to the required extent. One way that has been advocated is the idea of 'one-stop shops' in the district towns, where the farmers can individually or in co-operation acquire the services they need. The poor depend on functioning village level organizations to access such services. An important role of the extension staff is to strengthen the capacity of such village organizations which represent the interests of the poor.

Regarding advisory services, the poor tend to benefit more from communication within the village, rather than through contact with extension staff. Extension staff mainly target the better-off farmers who have the greatest potential for testing and developing new technologies. The ideas and technologies which may be relevant to the poor are then 'filtered' through the community organizations and discussed at village meetings, which are more accessible to the poor.

Mass organizations play a very important role in the context of discussing extension information at village level. These are primarily the women's union, the farmers' association, the youth union, the old people's union and the war veterans association. The women's union holds monthly group meetings for the exchange of knowledge and runs small-scale savings and credit schemes. The farmers' association organizes training courses, to which it invites extension organization staff or other relevant people. Membership fees in the mass organizations are low and are normally not perceived as a barrier to entry.

The mass organizations are more likely than the Extension Organization to involve the poor because of their mainly social objectives. They do not have production targets to live up to. Their success is measured by the number of people involved in their activities and how well anchored they are in the community. Their extension principles are based more on exchange of knowledge, rather than formal training courses, which is also more inclusive of the poor. Poor people may, nonetheless, still be left out, as there is often a connection between poverty and social exclusion. The situation is worse for the poor in better-off communities, where they form a relatively small percentage of the population, and are often excluded from services.

During the period of collective agriculture, before 1989, the co-operatives organized production for the whole village and people cultivated the land in brigades. When land reform was implemented in the early 1990s and tenure was transferred to individual households, the co-operatives collapsed in many areas. In the northern mountain area, there was a vacuum following the demise of the co-operatives, with all households having to arrange market and service contacts individually. This was difficult, as there were no widespread networks of private traders to replace the co-operative structures. The villages involved in the Vietnam–Sweden Forestry Co-operation Programme (FCP), 1991–5/MRDP programme[6] set up new village organizations, originally to manage project matters, but gradually and increasingly to fill the role of co-ordinators for farmer contacts and activities in extension, input supply, veterinary services, and revolving credit. The main way for them to finance their extension activities has been by combining extension with seed production services, input services, and marketing.

In other areas, the old co-operatives were dismantled and replaced by new ones, which adopted new regulations and sometimes elected a new leadership. There are now 6000 co-operatives, which follow the new law on co-operatives, and 60–70 per cent of them provide extension services. In some areas the co-operatives are still seen as top-down structures, inhibiting the development of private initiative. The co-operatives are sometimes criticized for not being real membership organizations, as most of the people in the village join almost automatically. However, this means that the poor are more likely to be included. The co-operatives tend to concentrate on economic activities of relevance to almost everyone in the community, such as water management (irrigation and drainage) and input supply (seed and fertilizer) for rice production. They distribute basic information on production, but their extension capacity tends to be limited.

## FIELD LEVEL STAFF

In total, the provincial extension centres have around 900 staff (15–20 per province), 70 per cent of whom have a university degree. At the district level, there are around 2000 staff. There is a general freeze on the employment of government staff, and directives for a reduction of 15 per cent by 2003. The provinces, which have not yet developed their extension system at the district level are thus not able to increase their staff unless staff are moved from other departments. It is up to the individual provinces to budget for and finance extension workers on contract at the commune level. Most provinces do not think that they can afford this. The northern mountain provinces have been able to use international donor funds and funds from Programme '135' to contract commune extension workers, who are more needed in the remote communes where there is less contact with district staff.

The salaries of government employees are generally below living standard requirements. Extension staff who have a university degree earn around 350,000 dong (US$ 24) per month. Apart from that, they may receive fieldwork allowances to cover the costs of transport and a midday meal. However, these allowances are not clearly regulated and are thus uncertain. The lack of adequate coverage of fieldwork costs is a major constraint in determining the outreach of staff. Extension staff are allowed to sign individual contracts with farmers for commercial provision of advisory services, but this practice is not yet widespread. Often contracts are not made individually, but with the extension organization.

Despite their low salaries, there is generally a high level of motivation among the extension staff, who often work overtime to respond to farmer demand outside the formal agenda. They are often admirable in handling the balance between state directives and local needs and circumstances. The local extension stations welcome support from development programmes in order to get resources so as to spend time on extension activities directed to local needs and the needs of poor households.

It makes a difference where the extension staff come from. Those who belong to the area that they are serving and who have grown up with its farming systems are more likely to give relevant advice. Social relations are also important. Extensionists working in their home areas come under social pressures, which increases their commitment and also the need to be realistic.

## FINANCING OF EXTENSION

The national government budget for extension programmes is steadily increasing and has grown from VND 14.3 billion in 1994 to VND 44 billion (around US$ 30 million) in 2001 (excluding staff and administrative costs). The budget is not very large by regional standards (Thailand's extension budget is US$ 150 million a year). Apart from the national budget, each province can also allocate funds for extension from provincial funds—ranging from 200 million VND (around US$ 13,000) in the northern mountain provinces to VND 1-2 billion (around US$ 650-130,000) in the southern lowland provinces (depending on how 'rich' the province is). This difference also reflects how extension inputs into lowland intensive agriculture are considered more profitable than those in remote mountain areas.

Resources for extension are allocated by the Department of Planning and Investment (DPI) at the province level, in competition with all the other provincial departments. The provincial level often focuses on fulfilling national goals and agendas. The district extension stations, which work directly with the farmers, more often want to concentrate on local priorities. In cases where these do not coincide with the national programmes, the district people's committee has to mobilize extra resources.

## OUTREACH

Public involvement in extension is generally focused on direct services to farmers. The outreach is, therefore, limited, as the number of extension staff is limited. Results have been better in areas where the extension staff concentrate more on capacity-building within local organizations with extension functions. The government realized at an early stage that local institutional development was crucial in the outreach of extension and access to extension by the poor. The MARD conference on extension in 1997 gathered representatives of a broad range of farmer groups and village organizations involved in extension at the commune and village levels, to exchange experiences on how to develop local-level extension and to draw policy conclusions.

There is widespread belief within the extension organization in the spread of knowledge and technologies through demonstration models and farmer examples. Theoretically, the idea is that farmers learn from their peers. Farmers are more likely to learn from farmers like themselves who practice a certain technology. In effect, the concept comes close to a trickle-down model. It is easier for the more successful (and, therefore, richer) farmers to access state funds for 'models', attract the attention of extension staff and receive credit. A consequence of quantitative production goals is that the extension organization concentrates time and resources on farmers with the best for production potential.

Although there is frequent mention of poverty alleviation in policy documents, conceptualizations of poverty are still anchored in perceptions and approaches from the 1980s, when a large percentage of the rural population was poor. In terms of current relative poverty within the communities, however, the focus of extension is seldom on the poorest. The poor tend to have less education and less knowledge about production. This is often used as the main argument for not including them in extension activities, since there would be less spread effect to other farmers and the rate of success with the new technology would be less.

There is an ongoing discussion about revising the policy directives for extension (Decree 13, 1993) and the need for increased poverty orientation in the extension system. However, the majority of policy makers still favour the demonstration model approach. They point to the fact that so far the overall growth in production and the rural economy has also benefited the poor.

The plant protection organization has the widest outreach and is relatively flexible in its response to farmer demand. This is due to the nature of its advice as a public good. Information on how to deal with outbreaks of pests spreads easily among farmers, who are keen to contain outbreaks. The plant protection organization also organizes training of private suppliers so that they are able to provide advisory services along with the sale of products. The staff of the plant protection organization have the most training in participatory extension approaches, because of the massive investment by the Food and Agriculture Organization (FAO) in building capacity for integrated pest management (IPM) training through farmer field schools. This training is widespread all over the country and has a big impact on farmers' knowledge of biological predators and the use of pesticides. Learning in farmer field schools is linked directly to practical experience in the field, and this increases the chance that the farmers really integrate the new experiences, thus also increasing the spread of knowledge and experience between farmers. When a large part of the community incorporates new knowledge in common agriculture practices, the poor are also more easily integrated.

The veterinary organization provides extension services in the area of vaccination campaigns. As animal health has an impact on the whole area and, therefore, also has characteristics of a public good, there is an incentive to try to include the poorest. This is, however, difficult and vaccination rates are often low among poor farmers due to the cost and scepticism regarding the benefits of vaccination. Village and commune veterinary services exist, but knowledge levels are often low. The state veterinary system plays a role in training and support for private veterinary workers and in certification and quality control for private stores of veterinary medicine. The poor get most of their advice regarding animal diseases from neighbours with some veterinary knowledge.

PARTICIPATORY EXTENSION

The MARD defines participatory extension[7] as the process whereby farmers and extension staff together analyse local conditions and needs, define activities, implement and evaluate them, and share their costs. The report concludes that the advantages are that resulting technologies are more appropriate to local conditions and less dependent on outside subsidies. The disadvantages are that the way of working has so far focused mainly on production and not attended to the needs of the whole household, especially women. It is stated that there is an overall need to improve mechanisms whereby the needs of farmers guide extension priorities. State directives on extension are acknowledged to insufficiently reflect farmers' needs.

There is demand for training of staff in extension methods, including participatory methods. A major contribution of the many international programmes in operation is that they provide opportunities for staff to try out new ways of working and build up new experiences, which they then adapt and take back into the government system. Many organizations focus mainly on capacity building and training of staff (see also Box 3.3). The World Bank project in the northern mountains supports the government network of training institutions, for technical, organizational and administrative skills, to enable commune and district level staff to be more active in managing development projects.

---

BOX 3.3

Donor-supported Initiatives: The Example of the Sida-supported
Mountain Rural Development Programme

The MRDP put great effort into supporting the development of local
institutional structures for extension and credit, supporting the villages
in organizing village institutions responsible for development planning
in the village, with the help of PRA methods. The villages then organized
their own extension and other development activities and handled re-
quests for services from district and provincial institutions. Many village
institutions now operate independently of project support, while others
faded out when the project withdrew, largely depending on the person-
alities and motivation of the village leaders, and on how institutionalized
the way of working has become at provincial and district levels. Some
provinces (such as Tuyen Quang) adapted a large part of their entire
extension activities to responding to demand from the village organizations. There is a close network of personal contacts between the extension staff and the village leaders and successful farmers. The extension
agenda set by this process is more relevant to the majority of farmers in
the villages than are the top–down programmes that dominate elsewhere.
Village institutions do not, however, automatically represent the specific
interests of the poor, especially if the poor are in the minority. Every-
body usually comes to meetings, but the agenda often tends to be set from
the point of view of the majority. There is a need for further analysis and
more attention to the specific situations and needs of the poor.

In 1996, the MRDP decided to shift the focus of the programme to
the mountain areas where the majority of people in the villages where
it operates are poor; and the programme could concentrate on development plans for the whole village. When the objective of reaching the
poor was brought to the forefront, however, the work with the development of participatory extension institutions in the midlands was
pushed to the background. Major achievements concerning institutions
for demand-driven services were thereby abandoned at a critical stage,
which significantly reduced the sustainability and institutionalization
of participation. Attention to the poor in the midland villages was
dropped, as was work on finding ways of responding to their demands.

---

## MAKING AGRICULTURAL EXTENSION AND RURAL DEVELOPMENT PRO-POOR: OPPORTUNITIES AND CONSTRAINTS

### OFFICIAL PERCEPTIONS

According to Decree 13 (1993), the role of the extension organization
is as follows: (i) to disseminate advanced technology in cultivation,

animal husbandry, forestry, fisheries, processing industry, storage, and post-harvest technology; (ii) to develop economic management skills and knowledge among farmers for effective business production; and (iii) to co-ordinate with other organizations in order to provide farmers with market and price information so that they can organize their production and business in an economically efficient way.

At present, the extension organization focuses mainly on the dissemination of technology for primary production, with 70–80 per cent of its funds being used for demonstration models. The other components of the extension agenda are limited by a lack of experience and capacity. The MARD policy documents recognize, however, that the extension organization is only eight years old, and that it is still developing and discovering appropriate roles and ways of working. There is also recognition that basic food security has been more or less achieved, at least in the lowland paddy areas. The focus of extension is thus to turn to broader, more livelihood related issues, such as business planning, efficient use of credit, market development, non-agricultural income generation, post-harvest technologies, and processing. Making this shift in the role of the extension organization will require a concentrating on the training of extension staff, both the currently active staff and the academic education of future staff.

THE ROLE OF RICE AND DIVERSIFICATION

Rice is still the basis of the rural economy; 70 per cent of Vietnamese households grow rice and 99.9 per cent consume rice. Rice accounts for three-quarters of the caloric intake of the average Vietnamese household. The rural poor in the lowland and midland provinces perceive an increase in rice productivity as the first priority for increased food security. A study by IFPRI (Minot and Goletti 2000) shows that rice consumption rises with increases in income for low and middle income households, indicating that food security in rice has not yet been attained. Rice, as a means to food security, is not questioned. Rice production for income generation, however, is becoming less attractive with decreasing profitability. Import restrictions on fertilizer are being removed, leading to higher prices for fertilizer and thus higher production costs while market prices for rice remain continuously low. The IFPRI study suggests that the removal of export quotas would raise prices by 14–22 per cent. There is thus a conflict of interest between the poor who are net consumers of rice and the very large group of farmers whose poverty status is largely determined by whether or not they can produce a marketable surplus from rice. According to the IFPRI study, the net effect in poverty reduction would be greater with market liberalization and higher rice prices. Another way would be to convert the export quotas to an export tax, which could then be used for redistribution purposes.

The increasing surplus in rice production during the 1990s has led to increased diversification, mainly into animal husbandry, but also into cash crops such as pepper, fruit trees, cinnamon, and coffee. Non-agricultural supplementary incomes are not yet widespread, but are highly desired. It is generally not seen as an exit option but as a supplement, which strengthens the household economy as a whole, including agriculture. Better-off households invest in rice milling, small tractors for ploughing, carpentry, tailoring, and other rural services.

Diversification is strongly encouraged in policy documents. Two main constraints remain, however, the lack of expertise among field staff concerning especially non-agricultural diversification options, and the difficulties in obtaining credit from the Bank for Agriculture and Rural Development for non-agriculture purposes.

Subsidies are often used to encourage diversification into new areas of production. State subsidies are provided to break traditional patterns, and provide extension and credit for trials. They are mostly linked to certain crops, such as beans, groundnuts and pepper, where seed is subsidized or distributed free. There are also funds managed by the state treasury, to which anyone can apply for subsidized credit for new production.

Trials for non-traditional production or niche products are going on at all levels, although with small resources. So far it has mainly been the better-off entrepreneurial farmers who are involved in niche production. Niche products are often market-sensitive and require high knowledge input. Diversification is important in order to spread risks, but spreading into new products is in itself also a risk. One niche product, which seems to be within the reach of the poor, is mushrooms, which requires minimal land holdings and can be grown with relatively low investment on rice straw. In Vinh Phuc province, mushroom production started a few years ago and is spreading fast. They have now established market linkages and are exporting to China. In Quang Tri province, the Department of Science and Technology has promised to buy mushrooms from farmers for marketing in Hanoi during a pilot period, until other marketing channels have been established.

## LIVESTOCK

Investments in animal husbandry increased enormously during the 1990s. Almost every rural household in the country has at least one pig and many have two or three. Pig-raising is and remains the most important source of supplementary income for the poor, and fulfils many functions: uses crop residues for fodder, provides manure, provides a means of saving. Pig-raising is especially important for rice farmers, to bridge the long income gap between the autumn and spring harvests. Pigs are sold mainly on the local market. Local organizations, especially the women's union, devote a lot of their effort to extension for pig-raising. The extension organization has focused on the introduction of new breeds with higher productivity. Crossbreed pigs for lean meat production are now widespread, even among the poor. New breeds of poultry are still at a trial stage, and the poor would do well to stay out of what are still clearly risky investments. Crossbreeds of cattle (Indian Sind) comprise one of the major extension campaigns, but they are not directly relevant to the poor, as they involve a relatively large investment.

Free grazed cattle production in the hilly areas has been important for the poor, as it requires very little continuous investment. Cattle also function as an important form of savings, which can be accessed in times of crisis or for major events such as weddings. Land for free grazing is becoming limited, however, and there is more competition with other land use, mainly forest planting. Intensive cattle-raising is encouraged from a veterinary point of view, because it is easier to keep diseases under control in stall-fed cattle. However, this limits the possibilities for the poor to keep cattle due to the reliance on production of cut-and-carry fodder and other factors.

## MARKET ORIENTATION

Vietnam has experienced exceptional growth in exports of agricultural produce during the 1990s. The agricultural economy has been transformed from an almost exclusive focus on subsistence and the domestic market, to becoming the second or third largest international exporter of a number of agricultural crops including rice, coffee, pepper, rubber, and cinnamon.

The boom in export production of crops other than rice has occurred largely in the midland and mountain areas. The expansion of cash crops, such as coffee, is the result of massive government campaigns, with the provision of land tenure certificates, credit, and input packages. Access to government resources has been relatively equitable, but the poor have suffered more from failures connected with lack of sufficient knowledge and production inputs. In some areas, there has been tension around access to land, as lowland farmers move in and gain access to land for export crop production at the expense of access for the ethnic minority population. Tension can also arise from export crop production dominating land use, at the expense of food production.

The rapid expansion of production has sometimes outpaced the development of marketing channels, information, and other components of the commodity chains. Incorrect market assessments have led to cases of large numbers of farmers not being able to sell their products. For example, the government's liaison with private processing industry in Thua Thien Hue province proved very risky, with the state providing massive resources to farmers for sugarcane production. The Singapore-owned processing factory decided to move to another province just before the harvest, leaving thousands of farmers without a market. Plum production in Lao Cai also expanded rapidly, out of phase with any processing industry. International market prices are currently falling drastically for coffee and pepper, resulting in serious losses for farmers. The Vietnam Coffee and Cocoa Association (VCCA) decided in 2001 that 30 per cent of the country's coffee trees should be chopped down (an equivalent of 180,000 ha of coffee), in an attempt to raise coffee prices.

The massive production campaigns are motivated by the need to achieve a sufficient scale of production to reduce transaction costs in relation to the remote areas. As each household produces on a small-scale, that is, from less than one hectare, the campaigns also result in households concentrating a large proportion of their resources on export crops, which makes them vulnerable to price reductions.

In the lowlands, people are gradually orienting themselves to domestic markets for vegetables and meat. The role of the government in this context is to build up the necessary regulations and infrastructure to ensure health standards and certification. Organically produced vegetables are in high demand, but an adequate certification system is not yet functioning. Organic production has so far mainly been an option in the market-integrated areas. There is significant competition in the domestic market from the countries in the region, mainly from Thailand and China.

Some export efforts have a chance of making a difference for the poor. Pepper is such a crop, since it requires relatively low inputs. Seedlings can be acquired from the plants of neighbours. It can be stored for a long time, so it is not as vulnerable to market fluctuations. The problem with using local seedlings is that the quality is lower and more liable to disease. The lower quality pepper is sold mainly on the local market. Vietnam is, however, rapidly increasing its export of pepper and is now the second largest exporter of pepper, after India.

The Dutch non-governmental organization (NGO) SNV, in Quang Tri and Hue, has developed a system of market development advisers for non-agricultural products. Groups of farmers analyse their situation and come up with ideas for income-generating activities. The advisers, who are local entrepreneurs, help with market contacts and linkages with people with useful experience who support and train the group to develop their ideas.

The CPRGS mentions measures to support the poor in their participation in markets; these include research in preservation methods and the provision of credit to reduce vulnerability to dips in market prices. The government will also, according to the document, help the poor to find suitable forms of economic co-operation to increase their bargaining power on the market.

LIVELIHOODS, COPING, AND VULNERABILITY

When a crisis affects a large part of the population in Vietnam, there is action by the government, local organizations, and people all over the country to mobilize resources to handle the immediate effects. For individual household crises, there is less support available. Safety nets are not well developed. Also, after a community-wide crisis, some people experience more difficulty in recovering than others. Households which were already indebted before the crisis and which have health problems in the family are among the most vulnerable.

Household coping strategies often include:

1. borrowing from family and friends—social capital networks are often strong;

2. borrowing money privately at 2–3 per cent interest per month;

3. borrowing rice at 40–50 per cent interest per season to be paid after the next harvest;

4. collecting minor forest products, such as leaves for hat-making and firewood;

5. seasonal migration to work as farm labourers; and

6. working locally as day labourers.

Even though people in the hilly areas are often poorer than those in the paddy-growing areas, the poor in the hilly areas tend to be less vulnerable to seasonal crises because of their more diversified sources of income, many of which become vital for coping strategies in relation to seasonal crises such as floods. The forest and communal areas provide important income during crises. In some areas, the forestry extension services support the enrichment of forests for household benefit and the development of minor forest products. Until recently, there were more employment opportunities available in the hilly areas, but they are becoming scarce with the crises in major export products.

Diversification can be seen both in the context of coping with crises such as these and in accumulation. Diversification by the hill land population after floods is clearly a coping strategy, as collecting minor forest products and taking day labour opportunities are poor people's means of diversification when other, more beneficial options no longer exist. These diversification options significantly reduce vulnerability to chronic indebtedness, which can lead to destitution. The better-off households diversify into non-land based production alternatives and business development, which reduces their vulnerability to both seasonal crises and market fluctuations.

Recovering from a major disaster, such as the floods in central Vietnam in 1999 or in the Mekong Delta in 2000, takes many years (Box 3.4). Incomes are not stable. In 2000, rice farmers in the central provinces incurred high costs for drainage because of the continued rains. In 2001, the harvest in the lower areas was lost because of flooding immediately before the harvest. The poor who are dependent on rice are becoming more and more vulnerable. They live in poorly built houses, which are more easily damaged in floods. Many people have suffered considerable losses because of repeated failures of investment in animal husbandry after the floods, when disease decimated livestock. Loans taken in connection with the floods are often impossible to repay, leading to spiralling indebtedness. In the

BOX 3.4
Diversification: Opportunities and Constraints

I (Dang) am 44 years old. We have three children. We have 1 ha of paddy, but no other agriculture land and only a small garden. In the big floods in 1999, we lost 1.2 tons of rice, one pig and 20 chickens. Another pig floated away, but was recovered alive. The house was damaged. We got food support enough for three months and we took flood recovery credit, VND 3 million which we used for rice inputs and two pigs and chickens. The pigs and chickens died after three months. Then we bought new pigs again. We don't know how to protect them if there are floods again. We have repaired the house, for VND 500,000.

We also used the flood recovery credit to pay interest on a private loan that we took three years ago to build this house, when we moved away from my parents house. The loan is VND 3.5 million with 2 per cent interest rate, which is 70,000 VND per month. We only manage to pay the interest and have not yet been able to pay any part of the principal. Household costs are 3 tons of rice per year and there is nothing left over to pay the debts.

In October 2000 we got a loan from the Bank for Agriculture of 3 million for animal husbandry, but we used it to pay back another private debt taken in 1999 to cover hospital and medical costs for our daughter. All three children have diseases related to the 'agent orange chemicals' in my body from the war, which make them lame if it is not treated. The oldest child can hardly walk. We have a 'poverty reduction card' which reduces costs by 50 per cent, but costs are still very high.

The harvest in 2000 was alright, we got 8 tons from both harvest and around 1300 VND per kg for the rice. But we didn't get any surplus because of the rains after planting which caused high drainage costs, and we had to replant twice. The spring harvest in 2001 was worse. It rained heavily in May, a week before the harvest. We only got 3 tons with low quality and only 1000 VND per kg. The autumn harvest was alright, but we are still in debt to the co-operative because of the bad spring harvest.

We don't have any other sources of income apart from the rice. We would like to try the new breeds of chicken, but don't dare to take the risk right now. There is not enough grazing and fodder for cattle raising. Animal husbandry in general is risky. Too many diseases, even if we vaccinate. My mother has sapodilla in her garden. We would like to have more fruit trees. My dream is to have employment, so that I have a secure regular income. Then we could do the farming as a side occupation and not be dependent on it (Vuong Khanh Dang and Nguyen Thi Suong, Phuoc Dien Village, Hai Thanh Commune, Hai Lang District, Quang Tri Province).

mountains, the difficulties in recovering from the floods are partly linked to lack of health and labour capacity to restore the land. The main damage in the mountains during the recent floods was caused by the enormous layers of stone and sand deposited by flood waters on fields close to the riverbed.

A challenge for extension is to incorporate an increased awareness of vulnerability. In the mountain areas people no longer feel safe in concentrating production on the flat land close to the rivers. There is increased pressure to find modes of production which make it possible to cultivate on the hill slopes again. In the lowlands, people need a diversification of sources of income to spread their risks and provide more varied means of recovery in the case of a crisis. New advisory priorities are needed to support indebted households that are rebuilding their economies and livelihoods. Advice on how to restructure loans and avoid high-interest private debts would have a significant impact on poverty reduction.

A common reason why poor people are excluded from services such as credit and extension is that they often use the resources 'for the wrong purposes', for example for housing, medical costs, or repaying old debts, and not for the production purpose intended. Policies aiming at reaching the poor need to take this into account. Purely production oriented services are often not feasible for the poor. More integrated forms of services are required.

A health crisis is a very common reason for people to become poor, and also for their difficulty in getting out of poverty. When a family member falls seriously ill, the family often has to sell off productive resources, such as a buffalo, to cover medical costs. These costs involve not only payment to the hospital, but also transport, lodging, and the cost of living in town. Sometimes bribes are involved just to access hospital services. The poor are entitled to reductions in hospital fees, but costs are still high. In addition, there is the cost of loss of labour capacity.

Community organizations are very important in dealing with all kinds of crises. The sense of community is strong in many places and the community often has mechanisms for supporting households which are in difficulty. The traditional village organization is often the most important socially, even though it has no material resources. The commune people's committee may have funds reserved for social support purposes. The mass organizations mobilize funds, and organize mutual support groups and groups for exchanging labour.

In the mountain communes, which are not so monetarized, informal lending of food and resources between relatives and neighbours is a common way of handling crises. Poor people in the lowland communes are more dependent on high-interest private loans to cope with crises.

The poor in the lowland and midland communities are often in a minority in their communes. This can put them in a more difficult situation than in the mountain communes, where the majority are poor and conditions are relatively similar. The poor minority often have less social capital in the community. This is also noticeable in terms of access to extension. They are less likely to be invited to training courses and extension activities. The possibilities of the poor in accessing services can be increased by investments in human capital, such as basic education and awareness of rights and opportunities. This is expected to also have an impact on the social integration of the poor in community organizations and in relations between the community and district and provincial services.

## Conclusions

International debate has focused mainly on aspects of extension as a vehicle for technical change and market orientation of production. When revisiting extension from the poverty perspective, we find that a broader approach is required. A focus on technologies alone cannot make a difference for the poor, who often need support in raising their general level of knowledge in order to access the information and services available and strengthen their bargaining power *vis-à-vis* service and input providers. More attention is needed on the possibilities of the poor in handling risk, including increased knowledge to avoid production failure, as well as insurance and safety net systems to make people less vulnerable.

An extension system, which provides specific technologies, can be organized on a commercial basis. A 'livelihood extension system', with the purpose of raising the level of knowledge and bargaining power of the poor, lies closer to basic education and cannot be regarded as a commercial service. The organization that is closest at hand for this purpose may not be the extension organization but the various community organizations.

We are thus talking about two different types of extension. Calls for the commercialization of extension services, cost-recovery, increased private sector involvement, etc. may all be relevant with regard to 'technology extension'. These extension services are in many cases not yet within the immediate scope of the poor. 'Livelihood extension', on the other hand, would focus public services on building the capacity of the poor to safeguard their interests in relation to market and service institutions.

Also, production-oriented services need to be viewed in the context of development of other institutions essential for poor people's livelihoods. Several such considerations are relevant.

1. Measures to reduce risk and vulnerability. The lack of appropriate safety nets and insurance systems is a significant constraint with respect to the possibilities of the poor developing their production and livelihood systems.

2. Policies regarding access to land in the mountains are a crucial issue affecting food security. Attention needs to be given to food security in the remote mountain areas, if people are to regain a willingness to confront the risks of market production. At present, food security is constrained by the lack of recognition of traditional farming systems and obstacles to access to land for food production.

3. The lack of access to services is connected both with physical remoteness and with social marginalization. Attention is concentrated on reducing physical remoteness, with investment in infrastructure, which is valuable. Less attention has been given to dealing with social marginalization, which is more common for the poor in market-integrated areas.

4. A large proportion of the rural poor are own-account farmers and need rural employment opportunities as a supplementary income (either labour or business) rather than full-time labour. There is need for capacity-building within the extension services to enable them to advise on off-farm production, business development, and micro-enterprise of various kinds. It is important that the poor are supported in their capacity as producers in this context, and not only as a source of labour.

5. The poor are often dependent on community organizations to access extension services. The public extension services should concentrate on building the capacity of local organizations, which would have a broader and more poverty-oriented effect than that achieved through the current concentration on model farmers.

## Endnotes

1. In this study we only discuss rural poverty.

2. Primary school enrolment rates have increased from 87–91 per cent for girls and 86–92 per cent for boys, in the period 1993–8. Lower secondary school enrolment has doubled and is now 62 per cent for both boys and girls. Upper secondary school enrolment has increased from 6–27 per cent for girls and from 8–30 per cent for boys.·In 1998, 58 per cent of the population owned a TV and 76 per cent owned a bicycle (World Bank 1999).

3. *Attacking Poverty* (World Bank 1999), drawn up for the Consultative Group meeting in December 1999.

4. It is usually referred to as Programme '135' for development in the 1000 communes 'in most difficulties'.

5. The Ministry of Labour, Invalids, and Social Affairs (MOLISA) is responsible for the Hunger Eradication and Poverty Reduction Programme and programme '120' for rural employment creation. The '135' Programme for the communes in most difficulties comes under the Ministry of Planning and Investment. The Ministry of Science, Technology, and Environment also plays an important part in rural development.

6. Vietnam–Sweden Forest Co-operation Programme 1991–5; Vietnam–Sweden Mountain Areas Rural Development Programme 1996–2001.

7. Draft policy guidelines, MARD 2000.

## References

Beckman, M., Le Van An, Le Quang Bao (2001), *Living with the floods, Coping and Adaptation Strategies of Households and Local Institutions in Central Vietnam*, Stockholm: Stockholm Environment Institute.

Nguyen, Dinh Huan (2001), 'Some lessons in the development of rural service enterprises', Unpublished, in Vietnamese language, Hanoi.

MARD (2000), 'Final Report, The extension activities 1993–2000 and orientation of extension activities 2001–2010', Draft, in Vietnamese language, Hanoi.

MARD and MRDP (1997), *The National Seminar on Agriculture and Forestry Extension*, Hanoi: MARD.

Minot, N. and F. Goletti (2000), *Rice Market Liberalization and Poverty in Vietnam*, Washington, DC: IFPRI.

MOLISA (2001), *Poverty Alleviation Strategy 2001–2010*, Hanoi: MOLISA.

Pham, Duc Tuan (2001), 'Consultancy report on agriculture and forestry extension', Unpublished, in Vietnamese language, Hanoi.

Pham, Duc Tuan and Thi Bang Nguyen (1997), *Extension and Rural Credit in Vietnam*, Hanoi: MARD.

Shanks, E., Bui Dinh Toai, Pham Dung Dai, and Vo Thanh Son (1999), *Participatory Poverty Assessment, Lao Cai Province*, Hanoi: MRDP.

Socialist Republic of Vietnam (2002), 'Comprehensive Poverty Reduction and Growth Strategy', Draft, Hanoi.

World Bank (2002), *2002 World Development Indicators*, CD Rom, Washington, DC: World Bank.

World Bank (1999a), *Vietnam—Voices of the Poor*, Hanoi: World Bank.

World Bank (1999b), *Vietnam Development Report 2000: Attacking Poverty*, Hanoi: World Bank.

World Bank, ADB, and UNDP (2000), *World Development Report 2001: Vietnam 2010: Entering the 21$^{st}$ Century*, Hanoi: World Bank.

# 4
# Extension, Poverty, and Vulnerability in Uganda
*Andrew D. Kidd*

## THE COUNTRY CONTEXT

### BASIC INDICATORS

Recent evidence from household surveys suggests that growth is reducing poverty across society. At the national level, however, there seems to be no systematic trend in the Gini coefficient, which gives an indication of inequality. Though the human immuno virus/ acquired immune deficiency syndrome (HIV/AIDS) pandemic remains of serious concern, Uganda is acclaimed for having begun to turn back the tide. It is widely recognized as one of the few countries in sub-Saharan Africa which is making progress towards sustained economic development and poverty reduction. However, most economic and social indicators remain weak (see Table 4.1).

### RURAL POVERTY

Uganda's population remains largely poor, with per capita gross domestic product (GDP) averaging only about US$ 330 and at least 40 per cent of the people living in poverty. A distribution map of income poverty shows that the north and east tend to be poorer than other areas (Map 4.1). These are areas where the seasonality of rainfall patterns is more pronounced and which suffer from greater insecurity. They are also more sparsely populated. The economy remains dependent on donor assistance and on the agricultural sector for both food self-sufficiency and foreign-exchange earnings from a variety of production systems. Many of the rural poor remain largely

TABLE 4.1
Uganda, Basic Indicators

| Series | Value | Year |
|---|---|---|
| Cereal yield (kg per hectare) | 1364 | 2000 |
| Land use, arable land (hectares per person) | 0.23 | 1999 |
| Land use, irrigated land (% of cropland) | 0.13 | 1999 |
| Agriculture, value added (% of GDP) | 42.5 | 2000 |
| GNI per capita, Atlas method (current US$) | 300 | 2000 |
| Population, total | 22,210,000 | 2000 |
| Rural population (% of total population) | 86 | 2000 |
| Malnutrition prevalence, height for age (% of children under 5) | 38 | 1995 |
| Malnutrition prevalence, weight for age (% of children under 5) | 25.5 | 1995 |
| Low birthweight babies (% of births) | Not available | – |
| Poverty headcount, national (% of population) | 55 | 1993 |
| Poverty headcount, rural (% of population) | Not available | – |
| GINI index | 37 | 1996 |
| Mortality rate, infant (per 1000 live births) | 83 | 2000 |
| School enrolment, primary (% net) | 87 | 1997 |
| Surface area (sq. km) | 241,040 | 2000 |
| Roads, total network (km) | 27,000 | 1999 |

Source: World Bank (2002).

outside the monetary economy, producing mainly for subsistence. Food crop production still accounts for at least 65 per cent of agricultural GDP, and agriculture continues to be characterized by low productivity. Challenges of rural economic transformation and poverty eradication linked to progress in the agricultural sector thus form the core of government policy.

While Uganda remains a very poor country, recent figures give encouragement.[1] The 2001 *Poverty Status Report* (MFPED 2001) notes that data from five nationally representative household surveys since 1992 suggest a consistent downward trend in income poverty, with the government on target to attain its income poverty reduction target of 10 per cent by 2017 (Figure 4.1). The proportion of people living below the income-based poverty line has declined from 56 per cent in 1992 to 35 per cent in 2000. Much of the progress in the

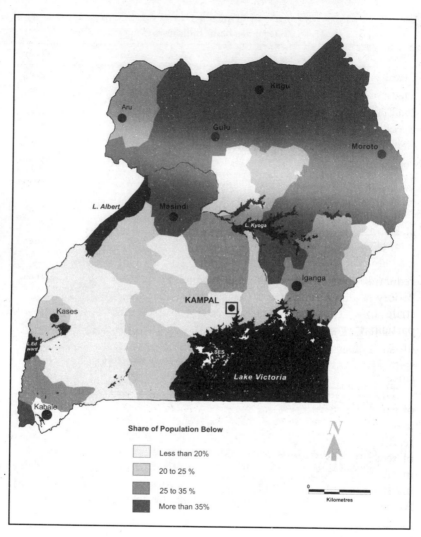

MAP 4.1: Spatial distribution of poverty in Uganda

*Source:* FEWS 1997, from USAID-funded Famine Early Warning System, adapted from World Bank, *Uganda: The Challenge of Growth and Poverty Reduction,* 1996.

*http://www.reliefweb.int/w/map.nsf/wByCLatest/ECF060A8E165815685256A0D006 FAC55?Opendocument*

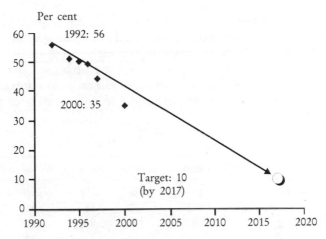

FIGURE 4.1: Poverty trends in Uganda

reduction of poverty can be linked to a successful rehabilitation process in many areas after years of political, economic, and civil strife. In Soroti and Katakwi districts, for example, Dutch-sponsored participatory rural appraisals (PRAs) used for local government planning were carried out in every sub-county in a rolling process over a 2–3 year period and showed a strong trend of resource accumulation since the mid-1990s.

The trends in poverty over the last ten years are taken as a source of encouragement for the aggressive reforming policies of the government and are used by international organizations and donors to reaffirm the reforming agenda in Africa. A 1999 World Bank review of sector performance indicates, for example, that liberalization has meant that producers have captured a larger proportion of the world market price for coffee—Uganda's largest export crop (Gibbon 2000). As mentioned above, however, the extent to which the trend in poverty reduction is sustainable is contested, given limited private investment flows, poor technological change, and over-reliance on raw agricultural commodity exports.

National data hide the growing regional and rural–urban inequality, with the south of the country, and particularly the central region, doing far better than the north. A regional analysis shows that the proportion of income poverty in rural areas is 26 per cent in the central region, 29 per cent in the western region, 39 per cent in the east and 67 per cent in the north.

This regional variation has an impact on the opportunities for households to reduce their poverty through, for example, wage labour. The price of wage labour varies significantly between the regions (Table 4.2), thus providing an incentive for migration particularly from the north (UBOS 2001). Non-agricultural wages are generally higher in most regions, with further disparity according to gender.

TABLE 4.2
Wage Rates per Day, 1999–2000 (Ush.)

|  | Central | Eastern | Northern | Western | Uganda |
|---|---|---|---|---|---|
| Agricultural, men | 1630 | 920 | 550 | 1020 | 1030 |
| Agricultural, women | 1420 | 860 | 550 | 940 | 940 |
| Non-agricultural, men | 1640 | 1230 | 820 | 1200 | 1280 |
| Non-agricultural, women | 1230 | 1130 | 540 | 1070 | 1020 |

*Note:* US$ 1 = 1600 Ush.
*Source:* UBOS 2001.

Policy documents also highlight gender disparity in the control of assets in favour of men, as also in the benefits from the sale of produce; though the extent of inequality varies with location. The Plan for the Modernization of Agriculture (PMA) takes the line that 'increases in household income do not necessarily mean increases in access to income for female members, or improved quality of life for all members, especially in terms of the nutrition of children'. In some instances, the growing interest in cash crops, controlled mostly by men, has meant that women have lost access to land for food crops (MFPED 2000b). While it is difficult to make definitive conclusions regarding the impact of gender disparity in relation to agricultural modernization and overall household benefit, it seems clear that there is a tendency towards greater household livelihood security when women have a greater share in the control and allocation of assets.

A programme of affirmative action by the government, which guarantees women a third of the total seats in local government, has led to 39 per cent of more than 17,000 strategic decision-making positions[2] in Uganda to be held by women (MGLSD and UBOS 2000, quoted in MFPED 2001). However, the access of women to non-mandatory, decision making positions remains poor (Table 4.3).

The Uganda Participatory Poverty Assessment Project (UPPAP) and other community studies (for example, COWI 2000) highlight

TABLE 4.3
Percentage of Women among Decision makers, 1999

| Administrative level | Major sector | | |
| --- | --- | --- | --- |
| | Political | Non-political | All sectors |
| Central government | 19 | 13 | 14 |
| Local government | 45 | 10 | 42 |
| All levels | 44 | 12 | 39 |

Source: MFPED 2001.

people's experience of households moving up or down the poverty-wealth scale. As some community members have highlighted, 'Yesterday's non-poor...today's poor' (Sebstad and Cohen 2000). Factors commonly noted as enabling households to move up the poverty–wealth ladder included business skills, access to finance, getting salaried employment, diversification of income sources, and obtaining remittances from children. The key features leading to increased poverty were identified as over-reliance on one source of income, human ill-health and disease, alcoholism and 'womanizing', and poor management of resources, including credit. By far the most critical factor determining movement in and out of poverty for many people was insecurity, with some one million refugees and internally displaced persons, particularly in the north of the country.

The dynamics of poverty in Uganda are largely determined by the various ways in which individuals, households, communities, and the national economy are vulnerable to external shocks, stresses, and trends (MFPED 2000b, 2001). The sources of vulnerability and the risks of remaining in or slipping into poverty are various; operate on a number of levels, and vary greatly across different parts of the country. These interact in various ways, potentially converging in a synergy of desperation.

Some of the poorest parts of the country suffer from a vulnerability cocktail that has human, political, natural, and economic elements compounding a growing regional inequality, which is itself a key cause of conflict (Stewart 1998, 2000). There is also some gender variation in the perceptions of risk and vulnerability, with women perceiving more risks associated with domestic factors than men (Sebstad and Cohen 2000). The importance of the various sources of vulnerability is hotly contested among stakeholders in Uganda. Some of the commonly cited causes of vulnerability are summarized in Box 4.1.

---

Box 4.1

Sources of Vulnerability

• climate change, uncertain weather patterns, seasonality, crop and livestock pests and diseases (for example, cassava mosaic virus), with particular impact in the drier northern and eastern regions;

• international trade regulations and globalization, with particular impact on the national economy and commodity producers and exporters (recent external shocks include the increase in oil prices, the crash in coffee price, the ban on imports of fish to the European Union, EU);[3]

• poor international market position due to higher transport costs, given that Uganda is a landlocked country; and perceptions of higher risk on the part of investors (for private investment flows), given the chronic political instability in some parts of the country, the wars in neighbouring countries and Uganda's direct involvement in the Congo war;

• rising prices, graduated tax burden, and seasonal variation in market prices;

• economic and social reform policies involving retrenchment and requiring users to pay for a variety of services (particularly health and education) that they find difficult to afford. Universal primary education (UPE), which has reduced the quality of public education through larger class sizes, has also made many vulnerable non-poor pay for private education, which puts them at greater risk;

• chronic political instability, insecurity, insurgency, cattle raiding, and child abduction in various parts of the country, and in terms of Uganda's relations with some neighbouring countries. The situation is exacerbated by the proliferation of small arms;

• the geographical isolation of some parts of the country, with high transport costs and poor market access;

• corruption and rent-seeking behaviour among public officials;[4]

• theft, community violence (a particular concern of men), and erosion of social cohesion;

• life-cycle events: births, marriages, and particularly deaths of family, friends and neighbours;

• sexual behaviour and the HIV/AIDS pandemic, which brings untimely death and has created large numbers of orphans, and child and female headed households;

• operating an enterprise as a result of the livelihood strategies chosen, particularly among the vulnerable non-poor with a limited skill base;

> - migration, with its risks and potential pay-offs: migrating workers may be preferred as they are willing to accept lower wages, poorer conditions, and have restricted choice and opportunity in the wage market, with less power and social recourse in their host society;
> - high dependency ratio in households and ill-health;
> - abandonment by spouse, domestic violence (faced by some 40 per cent of the women on a weekly basis in various parts of the country) and men's ownership and control of household assets, including land (concerns of women in particular).

The aspects of vulnerability that were commonly articulated in the recent participatory poverty assessment in nine districts (MFPED 2000b) include poor gender relations and other barriers to the effective participation of citizens in the governance processes, geographic isolation, insecurity at household, community, and regional levels, corruption and a lack of strong leadership, and cost-participation in service delivery (with a call for 'agricultural extension services that are accessible for all and offer relevant advice and information').

A strong message emerging from the UPPAP is that many rural people feel that their vulnerability is enhanced by the requirement for cost-sharing in service provision. This can have an influence on the livelihood strategies chosen. People's response clearly depends on their specific context in terms of the vulnerabilities they face, the mediating macro and micro policy environment, and the assets at their disposal. This varies not only from region to region and community to community, but also between individuals. However, a number of trends are discernible.

Diversification is common, with the poorer often showing diversification of individual strategies and the non-poor showing some diversification at the household level together with increasing specialization by individuals. The level of livelihood diversification is highly influenced by the perception of security, with significant withdrawal into subsistence activities in situations of chronic political instability. Remittances from migration can also be an important component of household livelihood strategies, though they seem insufficient to lift vulnerable households out of poverty (Mackinnon and Reinikka 2000).

In comparison with many other countries in sub-Saharan Africa, Uganda has agrarian systems that tend to be reasonably resilient, including to the impact of HIV/AIDS (Topouzis and du Guerny 1999;

Topouzis 2000). Much of the country is blessed with agro-ecosystems that are particularly well-adapted to diversification, that is, there is rain much of the year, or at least bimodal. This has clearly enabled many producers to follow a livelihood strategy which is diversified to some extent according to the 'rhythm of the vulnerability context'. Many rural people have production strategies for dealing with the rhythm of non-production related household expenditures. For example, in Lira district the sale of sesame is timely for the payment of second term school fees, and the maize harvest coincides well with the payments for the third term. The first term remains a struggle.

It seems that, while producers have tried to develop strategies to cope with these cyclical events, public policy may not have been sufficiently responsive to fit with the asset liquidity of clients. In a sense there is probably some scope for thinking more strategically about technical change in agriculture and the rhythm of rural livelihoods (for example, fresh and dry produce, determinate and indeterminate, and annual and perennial crops). Technology development and advisory services have rarely been able to incorporate a livelihood perspective. Of course, the unpredictability of various other events also shapes the choice of a livelihood strategy.

POLICIES TOWARDS AGRICULTURE AND RURAL DEVELOPMENT

Uganda is an interesting case for several reasons. The country is widely recognized as having made great strides in mainstreaming a poverty eradication agenda into government policy. This has shown encouraging signs of poverty reduction and may help achieve the millennium development goals. Furthermore, the notion of extension has recently undergone a radical shift, with the responsibility for extension being devolved to local governments, and contracting out being firmly on the agenda. Many of the developments are, therefore, in congruence with the common framework (Neuchâtel Group 1999; Beckman and Kidd 1999).

The Government of Uganda has been undertaking macroeconomic and development reforms since the National Resistance Movement (NRM) took power some 15 years ago, following a protracted constitutional crisis, highly centralized power, and civil war. The reforms are being implemented under a no-party democratic system in which significant public responsibility has been devolved to the districts, combined with a strongly forged programme for privatization and market liberalization. Dollar and Kraay (2001)—champions of globalization as a positive force against poverty—note that Uganda is among the countries they term 'globalizers', having substantially reduced tariffs on imports and increased trade in relation to GDP in the last twenty years.[5]

The strong economic reform programme underway since 1987 has the support of international financial institutions and donors. Liberalization, privatization, and decentralization are the key themes of the reform. The reform was undertaken rather reluctantly during the first five years of the process and was described as being crisis-driven (Tsikata 2000). From about 1992, the programme has been increasingly driven by the government itself and now has high levels of ownership (Holmgren et al. 1999; Mackinnon and Reinikka 2000). Holmgren et al. (1999) note that the government has made 'uncharacteristically good use of technical assistance'. The convergence of interest and perspective between the government and a broad coalition of national and international stakeholders has meant that the process of reform has not been as contentious as in many other countries of sub-Saharan Africa.

In response to the challenges of poverty and limited rural transformation, the government put in place the Poverty Eradication Action Plan (PEAP) in 1997. With some modification, this serves as a comprehensive development framework (CDF) and a revised summary of the PEAP was accepted as a Poverty Reduction Strategy Paper (PRSP). The revision was informed by the findings of the national household surveys, participatory poverty assessments, and a rolling stakeholder consultation process.[6] Uganda was the first country to benefit from debt relief under the highly indebted poor countries (HIPC) initiative, under which funds are channelled to targeted investment through the Poverty Action Fund (PAF).

One important modification in the PRSP is the greater emphasis on the private sector for an indirect impact on poverty. It affirms that the role of the public sector is to intervene where there is market failure or where there would be very inequitable outcomes, and prioritizes the use of contracting-out and development of partnerships with other sectors in society. The PRSP places distributional considerations—of gender, of children's rights, and of environmental impacts—at the heart of public policy (MFPED 2000a).

Ensuring good governance and security remains a complex and vital challenge for the government. The number of people affected by emergencies, both the drought in some parts of the country as well as the internal conflicts, has increased during the 1990s to more than 5 per cent of the population. Insecurity is probably the most important single reason for persistent poverty in the north, making serious development virtually impossible in some parts. International peace efforts, disarming the Karamajong, and dealing seriously with development issues are central to poverty eradication in these areas.

Despite being the backbone of the economy, the agricultural sector has seen serious under-investment, with only 1.8 per cent of the recurrent budget in 1998–9. A sector-wide approach, the PMA, aims to address this situation. The PMA is a bold statement of intent and a strategic vision for agricultural development, that will contribute to, as the full title goes on to suggest, 'eradicating poverty in Uganda'. It aims to be a 'holistic strategic framework for eradicating poverty through multi-sectoral interventions, enabling the people to improve their livelihoods in a sustainable manner' (GOU 2000). The document outlines a vision for 'poverty eradication through a profitable, competitive, sustainable, and dynamic agricultural and agro-industrial sector'. Levels of public action envisaged in the plan range from international trade negotiations to deepening decentralization and empowerment of producers. A legitimate role for a wide range of stake-holders in the governance of the agricultural sector is acknowledged in the PMA.

The plan notes that transforming subsistence agriculture requires addressing two types of constraints: productivity-related and governance-constraints related. The productivity-related constraints include a broad range from a lack of sufficient food, lack of land and soil infertility, lack of inputs, lack of skills and knowledge, lack of capital and access to credit and market problems (low prices, lack of markets), and poor roads and transport network. Insecurity is also highlighted as a productivity-related constraint with regard to the loss, for example, of oxen for draught power.

Within the PMA, public finance will be available, and this is intended to be targeted at providing services characterized as public goods. These include advisory services to enable producers to acquire knowledge and skills for agricultural transformation, product processing, and marketing. Support will also be given to smallholder-oriented agricultural research, capacity-building initiatives for rural institutions (including farmer organizations), rural market infrastructure, and regulatory services. The PMA document (GOU 2000) also notes what the government will not do, including supplying agricultural inputs, processing, or marketing agricultural outputs, subsidizing or providing credit directly to farmers; and constructing irrigation infrastructure.

Agricultural extension has been given a new lease of life in the PMA, which recognizes a central role for decentralized, demand-driven extension services in the sector's development and will be backed by public funds through the national agricultural advisory services (NAADS) programme. The responsibility for extension will be further delegated from districts to sub-counties, thereby deepening decentralization.

The government decided to devolve power in 1992, and the Local Government Act was enacted in 1997. The decentralization process involved substantial transfers of political, financial, and planning responsibilities, including agricultural extension from the centre to local governments. The Extension Directorate of MAAIF was abolished. The intention was to promote popular participation and the empowerment of local people in development planning and decision making, though participatory poverty assessments have shown that the intention has often been misunderstood as abandoning rural areas (MFPED 2000b). The accountability of service providers, including those offering extension services, was to be enhanced in the process, but many producers feel that this is far from becoming true. The NAADS programme has, somewhat controversially, taken the challenge of decentralization further by delegating responsibility from districts to sub-counties and broadening the role of producers in governance.

Most commentators agree that many of these structural changes have brought positive outcomes. The annual inflation rate has fallen from over 100 per cent in the 1980s to reach relative stability over the past few years. Over the same period, economic growth has come to average about 5 per cent per annum. This growth has been based on expansion in all sectors of the economy, with the agricultural sector (on which most of the poor depend) growing relatively fast, though much of the growth in agricultural GDP has been due to an expansion of the cultivated area rather than enhanced productivity.

Some commentators, however, question whether recent assessments have not been over-optimistic, suggesting that, for example, Uganda is better characterized as a 'stagnant low income country with poor development prospects' (Kappel 2001, pp. 23–4).[7] Bigsten (2000) suggests that recent trends are merely part of a 'rehabilitation process' following a dramatic decline in GDP per capita over two decades ago which began with nationalization, expulsion of Ugandan Asians, civil unrest, instability, plus the impact of the oil crisis. During this period, Uganda gained the reputation of being high-risk, contributing to a dramatic decline in investments and exports. Collier (1999a) highlights that the dividend from peace after a prolonged civil war is huge. Bigsten (2000) notes that recent growth is 'atypical, being more a dividend of policy reform and peace than, for instance, higher investment rates'. Collier (1999a) concurs with this view and suggests that recent economic trends reflect post-conflict recovery, the best example in Africa. Belshaw *et al.* (1999) argue that Uganda's recovery is much less impressive than it appears, since it is now a heavily aid-dependent economy. Moreover, the rehabilitation or recovery process is as yet incomplete, with several parts of the country still suffering from chronic political instability, which heightens regional inequality and leaves many vulnerable. Over 5 per cent of a population of some 21 million are internally displaced people (IDPs) and there is a significant number of refugees, in particular, from Sudan.

## AGRICULTURAL EXTENSION: BACKGROUND AND STATUS[8]

There has been some public financing of extension in one form or another since 1812. It is commonly considered that the extension service established by the government in the 1960s functioned well, but it declined during the economic and political crises of the 1970s and 1980s. The following chronology can be traced:

---

BOX 4.2
Agricultural Extension—Chronology

| | |
|---|---|
| 1812–1900 | Colonization and concentration on promotion of export crops. |
| 1920–56 | Extension through local chiefs, with enforced production of cash crops. |
| 1956–61 | Extension through progressive farmers; emphasis on provision inputs. |
| 1964–71 | Commodity approach with demonstration farms for transfer of technology. |
| 1971–92 | Political crisis, civil war. Disruption of the economy, centralization. Confusion. Some transition and recovery. |

| | |
|---|---|
| 1992–98 | Government Agricultural Extension Programme (AEP), with a 'unified extension approach' and the 'training & visit (T&V) system' introduced in phases in 27 districts. Criticizm of public extension services (example World Bank 1996). Various other bilateral financing arrangements and extension approaches. |
| 1998– | Village level participatory approach' (VLPA) introduced into the public extension service, the last death throes of the T&V system. Introduction of the graduate specialist scheme by central government, with responsibility for extension devolved to districts. Pluralism increasingly a reality. NGOs contracting public agents to deliver services, effectively privatizing the management of extension services in many areas. Support for advisory service deliv ery by farmer organizations through DANIDA supported Agricultural Sector Support Programme. |
| 1999–2001 | Finalization of the PMA, concentrating on food security through commercialization. Preparation for the NAADS programme, based on public finance, private delivery, contracting-out, demand-orientation, farmer-'ownership', cost-sharing, and decentralization to sub-counties. Basket financing arrangements supported by a number of donors. Support for advisory service delivery by decentralized farmer organizations. National Agricultural Research Organization (NARO) introduced the outreach programme. Various experiences with private sector development in service delivery, with support to advisory services for vertical integration and commodity systems approaches (for example, USAID funded 'Investment in Developing Export Agriculture' (IDEA) project of the Agribusiness Development Centre (ADC)). |
| 2001– | NAADS Bill passed by Parliament and NAADS Secretariat established as a corporate body. Phased introduction of the NAADS programme linked to broader decentralization of capacity-building initiatives, initially in six trailblazing districts (beginning with a couple of sub-counties in each district). Graduate specialist scheme to be phased out. |

Support to extension for the development of the agricultural sector has tended to focus on production, with limited attention being paid to wider livelihood contexts. This is likely to change with the introduction of the NAADS programme, which seeks to go beyond production and to look at supporting advisory services in marketing and processing.

## MAKING AGRICULTURAL EXTENSION AND RURAL DEVELOPMENT PRO-POOR: OPPORTUNITIES AND CONSTRAINTS

### OPPORTUNITIES FOR PRO-POOR DEVELOPMENT.

There are a number of areas that offer interesting insights on the scope of agricultural policy and extension to be pro-poor. The areas covered are not exhaustive and are intended to be illustrative of some of the key challenges and opportunities. They include: (i) organic produce; (ii) addressing market failure and transaction costs; (iii) addressing the digital divide; (iv) developing producer organizations; and (v) addressing chronic insecurity.

### Organic Produce

Conventional agricultural practices have had limited penetration and adoption in sub-Saharan Africa, and in Uganda in particular. Most agricultural production has been defined as being 'organic-by-default' or 'passively organic' due to the limited use of agrochemicals and the predominance of traditional production practices by the majority of smallholder farmers. From the perspective of commercial organic farming, less effort and lower investments are required for its conversion into 'organic' production. Organic certification in most circumstances in Uganda is a convenient ploy used by exporters to exploit a niche market and earn higher profits. It is in complete contrast to the image of organic farming in countries of the North, where farmers follow a substantially different philosophy and mode of production. In Uganda they need to do little more than simply signing a contract. Nevertheless, certification serves the same purpose and guarantees that the interests of the consumer are served.

Though organic produce has a very limited share of Uganda's export market (perhaps 1 per cent for coffee and cotton, and about 6–7 per cent for sesame exports), there are indications that some recent initiatives are tapping successfully into the growing demand for organic produce in the European Union, the main market, and that exports have the potential to rise significantly. These initiatives have been exclusively private sector-driven and, while following the spirit of the PMA, deviate from some of its production-related assumptions. An examination of the organic produce export market in Uganda highlights some interesting issues related to pro-poor opportunities that allow some beneficial impacts to be derived from globalization. There are also useful lessons about the patterns of interaction among different actors and agencies and interrelationship between regulatory and advisory functions.

The promotion of organic agriculture has had two main thrusts in Uganda. The first is based on income-generation objectives and is driven by the commercial private sector (with some 'market oriented' donor support); the second aims more at natural resource conservation and improvement in the livelihoods of resource-poor farmers, and is driven by NGOs (with some 'environmental and poverty oriented' donor support).

The first wave of formal promotion of organic agriculture occurred in the early 1990s when NGOs, with support from (mainly) European funders, adopted and initiated programmes under the label of 'sustainable agriculture', with the intention of arresting the degradation of smallholder farm lands and rejuvenating production to improve food security. While the production-oriented approach being promoted by NGOs has led to improvements in farm productivity, food security, and the incomes of the smallholder farmers involved, it tends to be expensive, with only a limited number of farmers benefiting (though there is an indication that some act as multipliers). The NGO approaches are knowledge-intensive; require significant training and advisory input; and include the promotion of gender equity, the organizational development of farmer groups, and, in some cases, other services such as revolving credit and the provision of initial organic inputs such as agroforestry tree seed(ling)s and equipment such as wheelbarrows and spades. There may also be some 'contribution in kind' expected from supporting donors. Most NGOs claim to target the most vulnerable members of the community, often referred to as the 'poorest of the poor', though this does not generally seem to be the case in local terms.

This production-oriented strand of support to organic agriculture takes place with little or no effort to bring the produce to the organic market. This can be attributed partly to a lack of marketing expertise among the NGOs, most of whom are production-oriented in their expertise, and to the lack of a local market for organic produce. The NGOs also have no contacts in the international organic market, however, some have recently started to support produce marketing and are exploring international organic marketing opportunities in collaboration with their European partners.[9]

The second wave to promote organic agriculture has tended to come from the 'for-profit', private sector, and is unreservedly market-oriented. In 1994, the Sida funded 'Export Promotion of Organic Produce from Africa' programme (EPOPA) began.[10] It was contracted out to the Dutch consultancy firm, Agro-Eco, and is being implemented in Uganda by its associate Agro-Eco (Uganda) Ltd. The aim of the project is to facilitate the sale of smallholder farmers' produce on the international organic market; this would provide a premium farm gate price, thereby increasing farmers' incomes. The underlying assumption is that 'farming practices in many places are organic and that the agriculture systems are very suitable for conversion to organic farming'. The driving force behind the programme is commercial and the main motivation of exporters and farmers to participate is the potential economic benefit to be derived from the higher prices obtained on the international market. The farmers tend to be paid an organic premium of about 20 per cent in comparison with that offered by the local conventional buyers (15–50 per cent).

There are a limited number of additional organic export arrangements. These are low-volume, high-value operations involving organic fruits (sweet bananas, pineapples, and avocados) and ginger. Exports are of the order of 2–3 tonnes per week. Under one of the arrangements, which is purely private with no donor support, the exporters (Bio Tropical Gardens Ltd and AMFRI Farm Ltd) have organized 10 to 20 farmers in a form of outgrower scheme around a pivotal entrepreneur. The farmers do not receive much technical input from the exporter, beyond information about what is prohibited in organic farming, though advice on quality and control of quality at the time of purchase is important. Again, the farmers have not had to change their farming practices much, if at all. The exporters cover the costs of certification (US$ 1000–1500 per year) and they, not the farmers, own the certification. In such cases, it has not been necessary to establish a field organization. Simple farm management records are adequate for the internal control system. The certification of such farmers is still based on farmer group certification and is the responsibility of the exporter. For such projects, 100 per cent inspection is possible and it is cost-effective, given the few farmers involved.

Many production systems in Uganda are based on a rich natural resource base with relatively good soils. A key opportunity for poorer farmers to benefit lies in the fact that many traditional production systems at present comply with the basic principles of the International Federation of Organic Agriculture Movements (IFOAM). In this case, being 'traditional' has a comparative advantage in the face of globalization. However, there are indications that many smallholder farming systems are under strain and some of the traditional husbandry practices, such as fallow, are no longer adequate to replenish soil fertility, given the shorter fallow cycles. In addition, some soils have limited supplies of phosphate and would require some external inputs for higher productivity.[11] It is certain that sustaining organic certification will inevitably require improvements in the farming systems, as degrading farms cannot be certified. Improvements in husbandry will require substantial extension inputs, which could be provided through the field organization and other agents. Yet the role of extension in the development of organic produce exports has been limited to date because the way farmers operate at present can often satisfy the requirements for certification. The need for training and advice is expected to increase as the sub-sector expands into areas where land is more limited or farming systems are in danger of becoming degraded. If the market share of organic exports continues to rise in Uganda, however, the advisory needs of the sector are likely to be under-supplied for both producers and exporters.

Expertise in organic agriculture is extremely limited in Uganda, with no training available at colleges and little explicit research into organic farming, though various specific technologies comply with it (Braun *et al.* 1997; Nakileza and Nsubuga 1999). There is great scope for further research into improved organic technologies, including improved agroforestry fallow technologies, rock phosphate, *rhizobium innoculum*, resistant crop varieties, and biological pest management. Clearly, technologies exist that can be adapted by local farmers to improve the productivity of their farming systems organically. The organic premium would also provide an incentive to adopt technologies that might otherwise have been characterized as only enhancing farm drudgery.

The type of training provided to public agents has not equipped them to address the demands of the sector properly. Much of the technical expertise lies with the various agencies in the NGO sector, which tend to draw on technologies developed outside Uganda and have shied away from developing relationships with the input-oriented public sector or the profit-oriented organic export entrepreneurs. This may change, given the potential convergence of some interests among the three groups[12] and the decentralized, stakeholder inclusive planning that will be strengthened through the NAADS programme. Competent NGOs, such as the Kulika Charitable Trust, have a local, intensive training input into the development of organic farming that could be tapped (and further supported) by farmer fora and linked with organic exporters.

Exporters also need advice when they begin to deal with organic produce. This may come from a number of sources: from certifiers, on regulations and procedures; from consultants, on the smooth and cost-effective implementation of an internal control system that complies with regulations and procedures; and from an intermediary organization (which may or may not also offer the consultancy above) or from business development services, for advice on marketing and market linkages. One such consultant, Agro-Eco (Uganda) Ltd, is today in a position to carry out feasibility studies into new organic projects that can then be used to encourage an exporter (with or without some donor assistance) to launch a new project, with some of its investment factored into the cost of any advisory work that is required during the project.

## Making Markets Work for the Poor

Making markets work for the poor is a central challenge if agricultural development is to be a poverty-reducing strategy, whether through direct (as producers) or indirect (as consumers or labourers) effects on the poor. Four strategic elements that require or facilitate a role for extension are highlighted here: (i) developing markets; (ii) linking poorer producers to markets; (iii) addressing the digital divide; and (iv) developing producers.

## Developing Markets

Agricultural policy (PMA) is quite clear about the need to develop markets and market-oriented agriculture among smallholder farmers in Uganda. A number of challenges remain, from international to local, which need to be tackled by a sector-wide approach.

Trade liberalization has been on the agenda during and since the Uruguay Round, and Uganda is making progress in complying with

its WTO membership commitments. With internal market liberalization in the early 1990s, the marketing boards for coffee, tea, and
cotton were dismantled, leading to higher farm-gate prices through
increased competition (Bigsten 2000; Bigsten and Kayizzi-Mugerwa
1999). The higher share of the gross price obtained by farmers
contributed to an expansion of production, which continued until the
collapse in the world coffee price after the mid-1990s.

Efforts are continuing to open up trade further and to capture more
benefits of globalization. Involvement in the evolving Common Market
for East and Southern Africa (COMESA), the inter-governmental East
African Co-operation and, the Cross-Border Initiative are likely to
bring about greater regional market integration. A number of international organizations[13] are keen to deliver trade-related technical
assistance via the implementation of an integrated framework to
capture the benefits of globalization.

Uganda's economy, however, is more vulnerable than those of
many other WTO members to trade trends and shocks, given that it
has a narrow export base—reliant on traditional export crops (coffee,
cotton, tea, and tobacco). Northern protectionizm, particularly for
the agricultural sectors in the EU and North America, adds to biases
in terms of trade against countries such as Uganda which are heavily
dependent on the export of agricultural commodities. As many of the
rules of the game are set externally, there is only a limited amount
that Uganda can do to address this situation. These factors, together
with high transport costs, have made it difficult for Uganda to make
a serious breakthrough in non-traditional export crops. The next
round of WTO negotiations on agriculture will be important, and it
is critical to develop Uganda's capacity in trade negotiations. Initiatives to boost policy advisory capacity will be important in this regard.

Linking poorer producers to markets will be one of the key
challenges in the coming years in Uganda. As already mentioned,
some agencies with a market orientation, whether through individual
farmers or producer organizations, prefer to work with more commercialized producers (probably ranging from vulnerable non-poor
to rich) who are considered to 'have potential'. This may well leave
a large number of farmers excluded, given that many poor households
cannot even produce enough food to feed their own families and are
net buyers of food, with food purchases commonly accounting for
60 per cent of monthly expenditure. This situation may lead to the
forced sale of assets in order to meet household needs, or to localized

casual labour and private food-for-work schemes that do not allow for asset accumulation. The situation of casual labourers is becoming progressively worse in relative terms, as the annual rise in the cost of casual labour is well below the rate of permanent or contract labourers (MFPED 1997).

There have been few initiatives to explicitly address this situation, which is likely to become an important feature of the NAADS programme. Many agencies have tended to address the lack of purchasing power for engaging in more commercial agriculture by providing handouts to poorer producers. This is increasingly being criticized for disrupting evolving exchange mechanisms in seed and inputs (see, for example, Longley 2001b). Alternative approaches may also allow subsidizing the acquisition of improved agricultural technologies, but in ways that reinforce rather than compete with local market mechanisms. Using vouchers (as was done by SG2000 and CRS, though with limited coverage) to transfer purchasing power to poorer producers for a private good (seed and fertilizer) can thus help address the excludability aspect associated with input-dependent advice and training that is otherwise being freely offered.[14]

The input voucher scheme organized by SG2000 is closely associated with the development of a rural stockist network, which can also fulfil an extension function by advising on and demonstrating the use of inputs to voucher recipients. A critical aspect of SG2000 has been to support partnerships between philanthropic organizations and the commercial private sector that do not lead to market failure. Thus, the agents of the programme (often NGOs and religious organizations) are able to target special interest groups (for example, the poorer, or female and child headed households, or those caring for AIDS orphans) without compromising (in fact, supporting) the development of a market-oriented rural stockist network. These organizations then alter their approach from input distribution (thereby bypassing important links in the market) to transfer of purchasing power (thereby supporting vertical integration of the market). Experience suggests that the schemes can operate well to address such issues related to input supply, but that they are costly,[15] open to the possibility of forgery, and require significant support and regulated community participation for effective targeting.

The agribusiness training and input network (ATAIN) component of the IDEA project provides an example of a more vertically integrated approach to advisory and other service needs among a range of actors in the agricultural sector. The project takes account of various links in the commodity chain for low-value, non-traditional exports, particularly by maize and beans. This requires strategic thinking about the constraints placed on the various actors involved in developing the sub-sector, especially input suppliers, distributors, stockists, extensionists, producers, and traders, and the patterns of interaction among them. Much attention in the past has been given to farmers' need for credit and advice about inputs. The project goes further by focusing attention on the information and advisory needs of support services, together with associated credit needs and risk management, and other links in the commodity system.

Other agencies are aiming to draw attention to other key actors who can help make things happen in market-oriented agricultural development. For example, FIT Uganda Ltd[16] has been developing a profile as an intermediary organization that can support training and business development services in the microenterprise sector. The SNV is also drawing on its positive experiences with FAIDA in Tanzania (see Christoplos et al. 2001) to develop a programme to address market linkages and support the role of business development services, financial services, and other intermediary organizations. The Co-operative League of the United States of America (CLUSA) may also become involved in Uganda and it will draw upon its experiences as an intermediary organization promoting market linkages and developing autonomous outgrower schemes that include farmer-managed extension and other service support.

When external assistance is necessary, developing market linkages and the role of intermediary actors will require different kinds of advisory services, particularly in view of the fact that there are advisory needs at different levels in commodity systems. The history of extension services and the way in which training institutions have developed in Uganda have primed them for increasing production in subsistence-oriented agriculture. They are largely unprepared for the challenges of developing a modernized agricultural sector.

There is also a need to get beyond the tendency for policy to suggest that richer farmers need specialist advice and poorer farmers can do with generalist advisory services for the sake of diversification. Reality is rarely so simple. Advice on diversification (farm management and enterprise development) is incorrectly thought to be synonymous with 'general advice', whereas in fact it is highly specialized. Producers of all kinds (and advisers themselves) are probably in most situations correct in their commonly felt characterization of 'specialist equals good', 'generalist equals bad', and this is one reason why the unified extension system, which operated in the mid-1990s with World Bank support, was hated by producers and advisers alike. It highlights the problem of using, for example, bland measures of coverage (such as farmer: extensionist ratios) as the main planning tool.

## Addressing the Digital Divide

Information and communication technology (ICT) can be an important aspect of support to rural development and the commercialization of agriculture, providing better information links for producers (or at least their organizations), traders, government, and service agencies alike. Various stakeholders involved in the agricultural sector can highlight the way in which the changing telecommunications landscape has transformed their work by improving access to information and contacts and reducing uncertainty and transaction costs associated with market development. Information and communication technology is also regarded as an avenue for reducing rural–urban migration, and for offering a more attractive rural future to young people. However, many of the poorer areas in Uganda remain outside these developments.

Compared with other African countries, Uganda's telecommunications market is highly competitive. The telecommunications industry has been privatized, and increasing competition has been reducing prices and improving quality since the Communications Act of 1997. Rapid developments have meant that the cost of transferring information over distances is falling dramatically in many areas and it is also becoming more reliable. The change in direct customer access has been particularly rapid in the last two years, with a tripling of subscriptions taking the total number of direct voice telephone customers to around 210,000. The privatization of Uganda Telecommunications Ltd (UTL) is likely to bring about aggressive growth. The second national operator MTN, which entered the market in 1998, has been able to secure more than 60 per cent of the market since it acquired a licence to provide both fixed and mobile services. There is also a competitively priced mobile operation (Mango), though it has only limited coverage.

While most telecommunications users are in and around Kampala, there is increasing rural access (Map 4.2), partly driven by market outreach and enforced by the licence agreement that requires mobile access in every district headquarter (some 44 rural districts) and at least one public payphone in each county (some 169 rural counties, made up of some 733 sub-counties and some 4020 parishes). The government has made rural outreach a condition of telecommunication licensing; in other words, it is using regulation to address aspects of market failure.

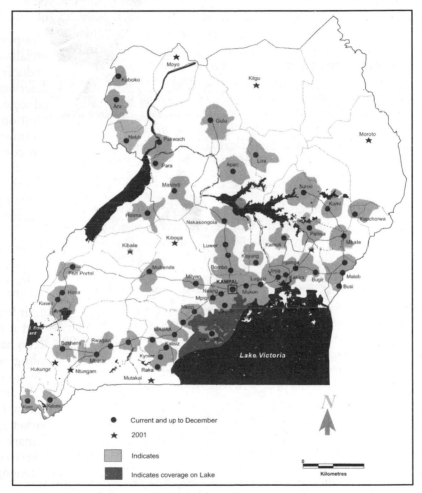

MAP 4.2: National operator MTN wireless telecommunication coverage
*Source:* Area of coverage denotes a 95 per cent chance of making a call. Coverage is subject to topographical conditions. *http://www.mtn.co.ug/products/coverage.htm*

The demand for telephone services is growing in rural areas, to the extent that would make service provision to much of the country commercially viable (according to estimates). Technologies also exist, for example, signal-enhancing, roof-mounted antennae and fixed wireless terminals together with solar power, that could make some localities in as many as two-thirds of sub-counties reachable from base stations by mid-2001. However, about 250 sub-counties have little chance of obtaining coverage by then and 150 of these will probably be unreachable even two years from now. Half of these will be in the poorer, more sparsely populated, and politically insecure northern and north-eastern areas of the country.

Most rural customers are not among the poorest, being mainly rural traders, headmasters, teachers, and middle to higher-ranking government and NGO workers. There is also an extensive informal market for services on the mobile network, where villagers pay handset owners on a call-by-call basis. Though the unit charge is higher, this access on demand is quite clearly preferred by most rural dwellers. The challenge that can partly be addressed through policy will be to improve access and reduce costs further. While the licence agreements of national operators providing one payphone per county have done something to address the access issue, the present and expected coverage remains clearly insufficient. It is widely acknowledged in the Ugandan telecommunications industry that more needs to be done.

The demand for ICTs such as email and Internet is much less, though some initiatives are looking at developing such capabilities. One key constraint is the lack of points of presence (POPs) among the six internet service providers (ISPs) capable of providing a similar level of service to that in Kampala, implying that long distance calls have to be made.

It is likely that the poor will benefit only indirectly from improved telecommunications via better-linked entrepreneurs, local governments, and service agencies. Agricultural entrepreneurs, buyers, and exporters have seen that their ability to reduce the risk associated with produce marketing is enhanced significantly with greater rural outreach of telecommunications.

There are indications that these changes are being used to support market development in the agricultural sector, partly through donor-financed initiatives and partly through private sector entrepreneurship (particularly with regard to market linkages and market information;

see, for example, Robbins and Ferris 2000). It is increasingly likely that the private sector will choose between similar areas on the basis of ease of communication, as a way of reducing risk and uncertainty. The logic of the PMA suggests that poverty reduction in poorer areas will require clear action in those areas which are technologically marginalized. At present, however, agricultural sector and extension policy lags behind in developments which address these emerging opportunities and challenges.

*Developing Producer Organizations*

The PMA highlights that capacity building for commodity associations, farmer organizations, and co-operatives will be critical during the early stages of the sector approach (GOU 2000). The NAADS programme will invest significant resources in farmer organizations to support the modernization of the sector. Self-governed and managed producer organizations of various types are widely regarded as important for the future development of the agricultural sector in Uganda, particularly for shaping policy, and in service provision and marketing. There is a feeling that agricultural marketing has been operating inefficiently following market liberalization, with too many middlemen and links in the marketing chain lowering the price obtained by the producers and distorting production incentives. Producer organizations are seen as an important but yet underdeveloped way of reducing the transaction costs involved in production and marketing. They can play an important role in enabling smallholder farmers to secure economies of scale in input purchasing and commodity trading. Various institutional arrangements and market exchanges can thus rely on group and social pressure or cohesion, rather than on the currently unenforceable legal contracts. This has proved useful in some situations of contract farming or outgrower schemes.

However, there are concerns that farmer groups and organizations in Uganda have not included the poor. More needs to be done to draw poorer farmers into co-operative arrangements (of various types) from which they can benefit by having greater economies of scale, more bargaining power, and greater voicing of their concerns. Some of those working within and supporting the district farmer associations suggest that farmer associations are not for every farmer, this view being rationalized in terms of working with producers with 'potential'. Given the unrepresentative membership, it is likely that many of the 'poor with potential' and 'vulnerable non-poor' are not included. The level of performance suggests that many farmers have made a rational choice not to become members (or in many cases not to renew their membership). Clearly, farmer associations will need to work at offering services demanded by farmers in the area.

There is some concern that the drive to sustainability and self-financing of farmer associations is unrealistically rushed. Farmers have been lukewarm about paying for the training and advice on offer from farmer associations; they are much more willing to cover the costs of input supply (when a lower price can be secured) or joint marketing (when a higher price is obtained). Farmer associations are expected to cover, in the first instance, 35 per cent of the cost of professional staff (for example, district co-ordinator, extension co-ordinator), with the share increasing over time. Most of the district associations staff receive only the 65 per cent secured from the donor, with no cost-sharing. Cost-sharing has, therefore, not met expectations, and the initial timeframe for attaining a certain level of membership and capacity is now regarded as unrealistic. Some associations are focusing their attention on developing activities that can generate income. Certainly, there seems to be a shift in understanding and expectation of the development of district farmer associations. Comments such as 'It took the Danes 150 years to get where they are today, 140 of which were government subsidized' are now more prominent than, for example, 'Farmers must pay for services that are primarily in the private interest' of a couple of years ago.

## Addressing Chronic Insecurity

Clearly much has been done to establish lasting peace in many parts of Uganda following the long period of civil war. The extent to which the economic turnaround can be sustained, or the peace dividend has been a matter of 'rehabilitation' or 'post-conflict recovery' rather than economic transformation, is open to question (Bigsten 2000; Collier 1999a). For many Ugandans chronic political insecurity, insurgency, and civil strife remain an important component of their livelihood context and drive the growing regional inequality, which, certain commentators (for instance, Stewart 2000; Addison et al. 2000; Luckham et al. 2001) warn as being a key determinant of conflict.

Beijuka (1999) highlights five conflict and post-conflict zones in Uganda (Map 4.3), some that were particularly affected by the civil war and others where there remains insecurity of varying degrees. In

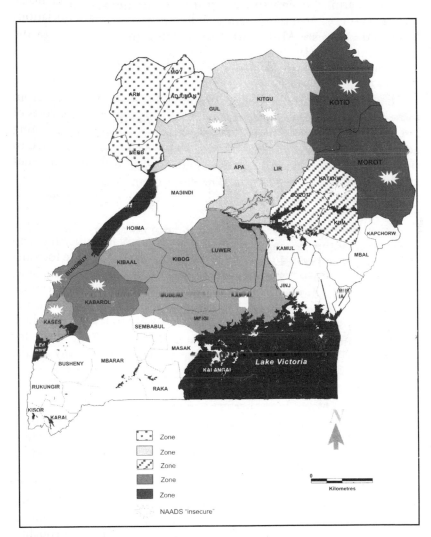

MAP 4.3: Conflict and post-conflict areas in Uganda

*Source*: adapted from Beijuka 1999.

*http://www.reliefweb.int/w/map.nsf/wByCLatest/0163A160F9C7F8CB85256A0D00*
*718D2F?Opendocument*

addition, the two districts that make up Karimoja (Moroto and Kotido) suffer from chronic insecurity, given the high numbers of small arms and the history of cattle raids. Northern Uganda is awash with small arms. Oxfam estimates that there are now about 160,000 guns in Karimoja, that is on average about one for every four persons. Most of the guns are AK-47s, which can be bought for about US$ 10 apiece near the Sudan border. This situation has deepened poverty, made many people highly vulnerable, and displaced around 1 million (Map 4.4).

The impact of chronic political insecurity on the livelihoods of those affected has been dramatic. Many have been killed or abducted,[17] lost livelihood assets, suffered personal trauma, or had their social networks and markets disrupted. The state has been unable to provide effective services in many areas, including agricultural research and extension, health, and education. The districts suffering from insecurity are among the poorest in Uganda and they have become increasingly marginalized.[18]

Civil strife reduces investment in non-farm enterprises, with subsistence agriculture becoming a larger and more resilient component of GDP (Collier 1999a; Deininger 2001, quoted in MFPED 2001; and Matovu and Stewart 2000). Luckham et al. (2001) highlight that civil strife in Uganda also led to a rapid decline in per capita agricultural production between the mid and late 1970s; since then it has stagnated, with a widespread return to subsistence farming.

The situation might well have been worse if market liberalization and attention to pluralism—allowing a greater role for non-state actors and agencies—had not become part of official policy. Before the conflict, most cash crops were marketed directly through localmarkets, whereas now there is a much greater role for middlemen, traders, and processors. In areas of chronic insecurity, the market is, therefore, still operating with clear inroads by the 'food security through commercialization' strategy promoted by the PMA (Longley 2001a).

Agricultural development will thus be an aspect of mitigation, and advisory and other support services will have an important role to play. As yet, however, these links have still to be clearly drawn in planning implementation of the PMA or the NAADS programme. The challenge will be to support the development of resilient production and marketing systems that can help overcome regional inequality and chronic insecurity.

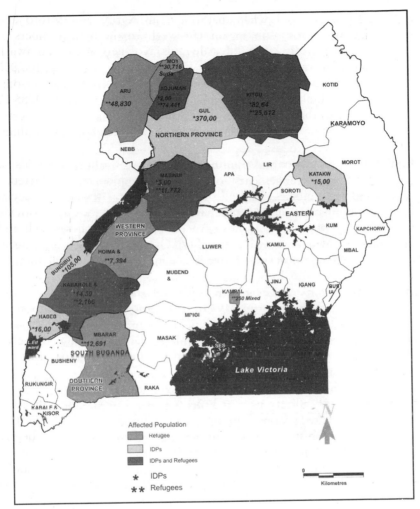

MAP 4.4: Distribution of displaced persons in Uganda, December 2000

*Note*: IDP—internally displaced person.

*Source: http://www.reliefweb.int/w/fullMaps_Af.nsf/luFullMap/24598BB75DB61*
*F83852569FA007DD367/$File/uganda_affected_dec00.pdf?OpenElement*
UN Office for the Coordination of Humanitarian Affairs, December 2000.

Longley (2001a, b) describes practical measures for moving from
relief to development in situations where farming is an option as a
livelihood strategy. She criticizes the tendency of many agencies to

distribute seeds and tools when operating in post-conflict and ongoing emergency situations. Looking at the seed system of the Acholi region, which is badly affected by chronic insecurity, she notes that a first step is to recognize people's ability to cope in the midst of chronic insecurity. These coping strategies can then be strengthened in a way that builds local networks of exchange mechanisms and markets, rather than facing economic disruption and competition from well-meaning humanitarian agencies. Three options are highlighted: (i) poverty-focused approaches that widen access to agricultural inputs; (ii) advice and training on alternative technologies; and (iii) institutional arrangements for access to/supply of agricultural inputs and technologies. Catholic Relief Services (CRS) have been using a combination of seed vouchers and seed fairs (incorporating an advisory service) as an innovative poverty-focused approach to assist farmers in insecure areas of northern Uganda. The seed vouchers transfer purchasing power to farmers which can then be exchanged at planting time with local traders during seed fairs, where there is also an opportunity to obtain advice on varietal choices.

Topouzis and du Guerny (1999) report that training and the provision of survival skills are seen as important aspects of helping orphans protect themselves from exploitation and abuse. They may also provide opportunities for positive changes in agricultural techniques, which would create links between advisory services associated with the wider livelihood context and agricultural productivity. There are indications that links between trauma counselling and market-oriented groups (who may also receive production-related advice) can have positive associations for building up social capital and reducing vulnerability (S. Lindorfer *pers. comm.*).

One vulnerability can lead to another. Luckham *et al.* (2001) note that the incidence of HIV/AIDS is accelerated by conflict. For many years, Uganda had the highest HIV infection rate in any African country. The infection peaked during the early 1990s, with an estimated 15 per cent of the population affected. This had an enormous impact, with many households slipping into poverty as a result of one or more members becoming infected or dying from the disease. Female-headed households in Uganda increased by 30 per cent during the 1990s, child-headed ones by some 40 per cent (MAAIF 2000a).

In the following decade, with peace and the adoption of appropriate prevention methods and programmes of mitigation, levels of infection have been halved. Uganda's National AIDS Control Programme included training community leaders, and mobilizing the community, innovative communication techniques to change attitudes, reducing discriminatory practices, and involving people living with AIDS in care and prevention activities.

At the household level, the impact of HIV/AIDS manifests itself in a loss of labour, leading to a decline in productivity and in household income, and a loss of assets, savings, and remittances, plus increased expenditure and an increased number of dependants relying on a small number of productive family members (Topouzis and du Guerny 1999). Most of the affected households are still largely dependent on agriculture, particularly for food for home consumption. The disease has had a particular impact on livelihood strategies that are highly labour-intensive such as agriculture.

Affected households are plunged further into poverty as they resort to selling off property in order to meet basic livelihood needs and medication. This leads to a decline in livestock production, which is often used as a form of saving. Topouzis and Hemrich (1996) showed that 65 per cent of AIDS-affected households were obliged to sell property to pay for care.

The disease has also hindered the effective operation of agricultural sector institutions and extension service organizations, with employee infection causing absenteeism, low productivity, job insecurity, low morale, and a lack of capacity to respond (MAAIF 2000). In MAAIF and the sector institutions, HIV/AIDS has more than doubled the expected number of deaths occurring among the workforce.

Support to affected households varies between promoting exclusively humanitarian assistance and engaging in developmental aspects. The provision of advisory services to vulnerable groups in relation to the pandemic has been a common approach, with advice ranging from trauma counselling to alternative productive activities. This has meant that there are a number of programmes that support the livelihood diversification of affected households, with support for micro-enterprises or agricultural enterprises that are less labour-intensive and have higher returns to labour.

OPPORTUNITIES FOR MAKING EXTENSION PRO-POOR

Poverty is multi-dimensional and the livelihoods of Uganda's poor people are often complex. The role that extension and advisory services of various kinds can play in improving livelihoods and contributing to poverty reduction is significant and will need to be considered at different levels and for alternative domains. There are links between changes in structures, policies, and processes and the ability, on the one hand, of governments to deliver a poverty-reducing framework and of poor people to effect sustained improvements in their livelihoods through enhanced opportunity, empowerment, and security. The recent history of Uganda in bringing about macroeconomic stability and poverty reduction demonstrates the connection. Yet Uganda remains a very poor country and more needs to be done, at the international, national, and local levels. Strengthening and broadening micro–macro links is one of the aims of introducing a sector-wide approach to development of the agricultural sector (the PMA) on which the majority of the poor depend. One key aspect will be to recognize the need for greater integration of advisory work in addressing different levels, from the needs of producers faced with an increasingly market-oriented economy to those of national stakeholders facing the challenges of globalization. The NAADS programme can make an important contribution though, on its own, this will be insufficient.

The poor in Uganda are producers, consumers, labourers, and citizens. Technical change in and around the agriculture sector will affect them in different ways, according to where they live and what livelihood strategies they choose. For many people, the indirect effects of agricultural development will be as important as the more direct production and market-related impacts. The Ugandan case highlights the fact that the position of the poor can be strengthened by three overlapping categories of action where extension, training and advisory services of various kinds can make, and are making, an important contribution: (i) creating and supporting opportunity; (ii) enabling empowerment; and (iii) improving security and reducing vulnerability.

## Creating and Supporting Opportunity

Creating opportunities will often involve local as well as international considerations. Within this, extension policy will need to be nested in a sector-wide strategy that builds a broad framework. The PMA is a bold step and should lay a foundation for action. The actual and potential elements of the strategy include:

1. strengthening inclusive policy processes, and building up capacities for policy analysis and negotiation in the international arena;

2. building institutions for inclusive markets (locally and nationally within a globalized economy), and developing the role of intermediary organizations in pro-poor market linkage arrangements;

3. developing vertically integrated strategies on commodity systems, and negotiating for, and supporting the capturing of, higher-value links in commodity chains within national boundaries;

4. instituting vertical integration or joined-up thinking on advisory services (for producers, stockists, traders, processors, and packagers and not only about producers producing), and with various links in the commodity chain where advice is best and most efficiently applied;

5. expanding rural connectivity, through targeted subsidies and regulatory mechanisms;

6. supporting access to a range of pro-poor micro-finance products (savings, credit, remittances), in addition to informal mechanisms;

7. broadening choice beyond low-risk, low-return enterprises for the poor;

8. building on strengths—taking a serious look at sub-sectors where the poor have comparative advantage, for example in organic production;

9. emphasizing quality through training and advice, supported by an appropriate mix of price and regulatory mechanisms;

10. introducing labour saving in smallholder production systems and labour intensification in larger-scale production systems; and

11. developing targeting mechanisms that strengthen, rather than compete with, local exchange mechanisms.

## Enabling Empowerment

Uganda has demonstrated an ability to construct greater political inclusivity (going beyond democracy). Decentralization and the UPPAP-based policy processes will strengthen this process. Decentralization itself will not necessarily be pro-poor (Moore and Putzel 1999), but it does provide an opportunity to address horizontal inequalities from which conflict tends to emerge (Stewart 2000; Luckham et al. 2001). Many of the policies to improve governance will benefit the poor. The mainstreaming of participatory poverty assessments in district planning will be an important step in this regard. Initiatives to strengthen producer organizations and the

NAADS programme are also important steps to place more power in the hands of producers. Sensitive implementation of the NAADS programme in the construction of farmer groups and fora can also open up opportunities for channelling the voice of the poor in agricultural development. The risk is that administrative mechanisms for the NAADS that depend on social capital may not act in favour of those who have least access to social capital. In addition, consideration must increasingly be given to how Uganda can itself be empowered within the globalized economy. Elements enabling greater empowerment, within which advisory functions on different levels will be of great importance, include:

1. building coalitions and negotiating capacity among countries similarly affected by rule-based changes in the global economy;

2. supporting decentralized capacity-building and the development of local government, with an important role for a cadre of non-sectoral advisers;

3. supporting intermediary organizations for better pro-poor, market-oriented institutional arrangements;

4. developing a broader range of support and advisory services;

5. building effective competition that favours maximization of producer prices;

6. introducing organizations and technologies to enhance bargaining power and reduce inefficient links in commodity chains, cutting out the middlemen;

7. valuing and building on the local mechanisms of support and exchange;

8. providing choice in technologies and service arrangements, and developing knowledge and skills across a number of domains; and

9. introducing institutions and administrative mechanisms to draw upon the voice of the poor in decentralized arrangements, ensuring inclusivity and enhancing accountability.

## Improving Security and Reducing Vulnerability

Technical change in agriculture and advisory services will contribute to reducing vulnerability and addressing chronic insecurity. Longley (2001b) highlights a critical role for advisory services in Uganda in addressing chronic political instability. Without tangible improvements in the livelihoods of people in the poorer regions, the cycle of conflict will be reinforced. A military solution is insufficient. The security and development of the poorer parts of Uganda will need to be increasingly addressed through public action (on various levels from the local to the international). Uganda itself is highly vulnerable, given Northern protectionism and the country's narrow export base. In order to benefit further from globalization, Bigsten (2000) notes that it is essential for Uganda to become internationally competitive in areas other than traditional commodity exports. Measures to achieve this include:

1. changing the rules of the game in international trade that act against the interests of poorer agricultural economies, and developing non-traditional export commodities;

2. addressing corruption and building confidence for greater private investment;

3. enabling poorer producers to go beyond low-risk, low-return production systems while building resilient systems;

4. providing greater opportunities additive rather than substitutive for more formalized savings and insurance mechanisms, without compromising informal coping strategies;

5. encouraging the blurring of the urban–rural divide by supporting financial services for the safe and efficient transmission of remittances;

6. implementing of the Land act, including the amendment in favour of vulnerable groups;

7. halting the trade in small arms, with a need for innovative policies to bring about peace;

8. training and providing survival skills to help, for example, orphans and women protect themselves against exploitation and abuse, and building links and synergy between, for example, trauma counselling and market-oriented advice to groups; and

9. recognizing and supporting the functioning of legitimate private sector initiatives and markets in areas considered insecure.

## Securing the Future

Uganda has made substantial progress in recent years in reducing poverty. Concerns remain that the apparent progress was a dividend of peace, and more a rehabilitation than a development process, thus raising questions of sustainability—given the limited private invest-ment and the low rates of technological change. Short to mid-term growth may benefit from Uganda's head start in policy reform, but long-term growth must seek out comparative advantage and technical change to draw in private investment flows. This will also need to be accompanied by changes in the nature of world trade so that the position of poor countries is strengthened.

This case study has highlighted some initiatives that address the issue of creating and supporting opportunity for producers in Uganda. But two key risks remain: the chronic insecurity in poorer parts of the country, and the vulnerable position of the national economy in the face of globalization. Technical change in the agricultural sector and advisory services of various kinds will make a contribution in bringing both direct and indirect benefits to the poor in Uganda. Chronic insecurity will require a positive programme to address the growing inequality in the country, and public investment will need to reflect such an imperative, certainly more than it does at present. This must go well beyond a military solution.

Furthermore, the PMA puts its faith in market mechanisms for eradicating poverty in Uganda. Reliable domestic and world markets for agricultural products are rightly regarded as crucial to success. The capacity for policy analysis and negotiation on global trade in regional and international fora, especially the WTO, is as critical as Uganda's ability to produce high quality products that are competitive in the global market. These capacities are at present underdeveloped within Uganda, which puts at risk its ability to maximize benefits and minimize losses that might arise from globalization. Poorer countries such as Uganda are only able to capture the lower-value links in the commodity chains, partly due to corporate interest and power and partly because of regulatory frameworks. These are critical elements of the national vulnerability context and may need action on the global rules of the game if poverty is to be eradicated. Bilateral donors will have an important role to play here and the type of donor convergence seen at the country level in Uganda will also need to be reflected in concerted action in the international arena.

### Endnotes

1. Some of the poorest districts, however, were not included in the most recent national household survey because of insecurity (Gulu, Kitgum, Kasese, and Bundibugyo) (UBOS 2001).

2. The definition of decision makers here was limited to the public sector, including the executive, legislature/parliamentarians, the judiciary, the public service, local governments, in addition to any other bodies that are decision makers as prescribed in the Ugandan Constitution (MFPED 2001).

3. The impact of the decline in coffee price was felt particularly strongly in Uganda, as in comparison with other coffee exporting countries between 1994 and 1998; it was characterized as a 'loser in a declining market', doing worse than the norm. Both slow liberalization of producer prices and failure to overcome early institutional resistance to marketing reforms contributed to this situation (Belshaw *et al.* 1999). The ban of fish imports to the EU was also a particularly strong shock to the economy and the livelihoods of those involved, given that during 1994–8 the fresh fish export sector had been regarded as an export 'champion' doing better than the norm (Elfring 2001).

4. Azam (2001) notes that a study of African countries shows a strong correlation between high public sector wages (often at the cost of the total number of civil servants) and peace in society. Problems such as the lack of a living wage among public servants during political, economic, and civil disturbance saw a consequent rise of alternative livelihood strategies among many public servants towards rent-seeking behaviour. For example, de Coninck (1992) notes that a bunch of matooke (staple starch bananas) sufficient for a family of four for 3 days costs Ush1000 in Kampala. On this basis a newly recruited messenger's salary would last five days and the monthly salary of the head of the civil service would last only 19 days. Uganda ranks 80th out of 90 in Transparency International's Corruption Perception Index 2000 (Transparency International 2000), sending signals that discourage private direct investment.

5. Uganda was arbitrarily added to the list of 'globalizers' along with Ghana, given that these two countries were close to the threshold and that the list would have otherwise lacked African representation. 'Globalizers' also include, for example, China, India, Mexico, Vietnam, Bangladesh, and Nepal. It should be noted that transport costs for a landlocked country such as Uganda typically add some 80 per cent to the costs of traded items (Morrissey 2000).

6. The UPPAP reported that people are frustrated by their perceived lack of influence over public policy. The UPPAP has itself helped to raise the voice of the poor.

7. This view, part of a broader analysis of development prospects for Africa by German researchers, has been recently described as simplistic and counterproductive (Hansohm and Thomas 2001).

8. For more insights on the status of extension in a particular district context (Mbarara) see Beckman and Kidd (1999).

9. TEFU and Kulika Charitable Trust.

10. A regional programme in Uganda, Zimbabwe, Tanzania, and Mozambique.

11. Some natural sources of phosphate are acceptable for use as external inputs in organic farming. The use of inorganic fertilizer would certainly be much more convenient and cost-effective in many cases of phosphate deficiency.

12. Interests such as environmental protection, enhanced commercialization and trade, and poverty reduction. There are clearly public good considerations that would justify some public investment.

13. These include IMF, ITC, UNCTAD, UNDP, World Bank, and WTO.

14. It is worth noting here that vouchers are only closely linked to 'demand' and 'accountability' when a farmer has choice. If there is little or no choice then they are merely a mechanism for transferring purchasing power.

15. There are extra operational costs in training operators, printing and handling vouchers, identifying beneficiaries, and in the additional profit margin to stockists and dealers for operating the system.

16. FIT Uganda Ltd is the local facilitator of the ILO FIT Programme supporting training, advisory and other business development services in the MSE sector.

17. Around 5000 children were abducted by the LRA between 1990 and 2000, with some 42 per cent yet to return. Child abduction is common in other areas of Uganda suffering from chronic insecurity.

18. The quality of information on poverty in some of the poorest districts is influenced by the fact that the 1999–2000. National household survey covered all districts except those most badly affected by rebel activity and insurgency—Kitgum, Gulu, Kasese, and Bundibugyo (UBOS 2001).

## References

Addison, A., P. Le Billon, and S. Mansoob Murshed (2000), 'On the Economic Motivation for Conflict in Africa', Paper prepared for the Annual World Bank Conference on Development Economics, Paris, June, *http://www.worldbank.org/research/abcde/eu_2000/pdffiles/murshed.pdf*

Azam, J. P. (2001), *The Redistributive State and Conflicts in Africa*, Centre for the Study of African Economies Working Paper Series WPS/2001.3, Oxford: Centre for the Study of African Economies, Oxford University.

Beckman, M. and A. D. Kidd (1999), *Reviewing the Neuchâtel Initiative Vision in Action: A case study from Uganda*, Report produced on behalf of the Neuchâtel Group with the support of Sida, Stockholm and GTZ, Eschborn.

Beijuka, J. (1999), 'Microfinance in Post-Conflict Countries: The Case Study of Uganda', Paper prepared for the Joint ILO/UNHCR Workshop: Microfinance in Post-Conflict Countries, ILO, Geneva, 15–17 September, *http://www.ilo.org/public/english/employment/finance/papers/uganda.htm*

Belshaw, D. and M. Malinga (1999), *The Kalashnikov Economies of the Eastern Sahel: Cumulative or Cyclical Differentiation between Nomadic Pastoralists*, London: Development Studies Association, South Bank University.

Belshaw, D., P. Lawrence, and M. Hubbard (1999), 'Agricultural Tradables and Economic Recovery in Uganda: The limitations of structural adjustment in practice', *World Development*, Vol. 27, No. 4, pp. 673–90.

Bennell, P., E. Masunungure, N. Ng'ethe, and G. Wilson (2000), *Improving Policy Analysis and Management for Poverty Reduction in sub-Saharan Africa: Creating an effective learning community*, Brighton: Institute of Development Studies at the University of Sussex.

Bigsten, A. (2000), 'Globalization and Income Inequality in Uganda', Paper presented at Poverty and Income Inequality in Developing Countries: A policy dialogue on the effects of Globalization' Conference, OECD Development Centre, Paris, 30 November–1 December, *http://www. oecd.org/dev/ENGLISH/pagelisteE/Poverty-Ineq/Documents/Poverty-Uganda.pdf*

Bigsten, A. and S. Kayizzi-Mugerwa (1999), *Crisis, Adjustment and Growth in Uganda: A study of adaptation in an African economy*, Basingstoke: Macmillan.

Botchwey, K., P. Collier, J. W. Gunning, and K. Hamada (1998), *Report of the Group of Independent Persons Appointed to Conduct an Evaluation of Certain Aspects of the Enhanced Structural Adjustment Facility*, Washington, DC: IMF.

Boyd, C., C. Turton, N. Hatibu, H. F. Mahoo, E. Lazaro, F. B. Rwehumbiza, P. Okubal, and M. Makumbi (2000), 'The Contribution of Soil and Water Conservation to Sustainable Livelihoods in Semi-Arid Areas of Sub-Saharan Africa', *AgREN Network Paper* No. 102, London: Agricultural Research and Extension Network, Overseas Development Institute.

Boyd, C. and T. Slaymaker (2000), *Re-Examining The 'More People Less Erosion' Hypothesis: Special Case Or Wider Trend? ODI Natural Resources Perspectives*, No. 63, London: Overseas Development Institute.

Braun, A. R., E. M. A. Smaling, E. L. Muchugu, K. D. Shepherd, and J. D. Corbett (eds) (1997), *Maintenance and Improvement of Soil Productivity in the Highlands of Ethiopia, Kenya, Madagascar and Uganda: An inventory of spatial and non-spatial survey and research data on natural resources and land productivity*, AHI Technical Report Series No. 6, African Highland Initiative, Nairobi: ICRAF.

Christoplos, I., J. Farrington, and A. D. Kidd (2001), *Extension, Poverty and Vulnerability: Inception Report of a Study for the Neuchâtel Initiative*, ODI Working Paper No. 144, London: Overseas Development Institute.

Collier, P. (1999a), 'On the Economic Consequences of Civil War', *Oxford Economic Papers*, Vol. 51, pp. 168–83.

———— (1999b), *The Challenge of Ugandan Reconstruction, 1986–1998*, Washington, DC: World Bank, *http://www.worldbank.org/research/ conflict/papers/uganda.pdf*

COWI (2000), 'Market Needs Assessment for Rural Financial Services in Selected Districts of Uganda', COWI Consulting Engineers and Planners

AS in collaboration with Nordic Consulting Group (U), Final Report for the Agriculture Sector Programme Support, May.

Dahms, M. (1999), *For the Educated People only…Reflections on a Visit to two Multipurpose Community Telecentres in Uganda*, Ottawa: IDRC, *http://www.idrc.ca/telecentre/evaluation/nn/14_For.html*

De Coninck, J. (1992), *Evaluating the Impact of NGOs in Rural Poverty Alleviation: Uganda Country Study*, Working Paper 51, London: Overseas Development Institute.

Deininger, K. (2001), 'Household Level Change in Uganda's Agricultural Sector, 1992 to 2000: Accomplishments and challenges', Washington, DC: World Bank Development Research Group, 22 January (mimeo).

Deininger, K. and J. Okidi (1999), '*Capital Market Access, Factor Demand, and Agricultural Development in Rural Areas of Developing Countries: The case of Uganda*', EPRC Research Series No. 12, Kampala: Economic Policy Research Centre, Makerere University.

Dercon, S. (2000), *Income Risk, Coping Strategies and Safety Nets*, Working Paper Series WPS/2000.26, Oxford: Centre for the Study of African Economies.

Dollar, D. and A. Kraay (2001), *Trade, Growth and Poverty*, Washington, DC: World Bank Development Research Group, *http://www.worldbank.org/research/growth/pdfiles/Trade5.pdf*

Elfring, W. (2001), 'Study on Agricultural and Rural Development Strategy for EAC', East African Community Secretariat, draft final synthesis report, February.

Food Rights Alliance (2000), 'The Plan for the Modernization of Agriculture: What does civil society think?', Unpublished manuscript.

Forss, K. and E. Sterky (2000), *Export Promotion of Organic Products from Africa: An evaluation of EPOPA*, Sida Evaluation 00/23, Stockholm: Department for Infrastructure and Economic Co-operation, Sida, *http://www.sida.se/evaluation*

Fuchs, R. and M. Mayanja (2000), 'Case Study No. 3: Multi-Purpose Community Telecentre, Nakaseke, Uganda', NTCA, pp. 66–80, *http://www.ntca.org/intl/FINAL.pdf*

Gibbon, P. (2000), 'The Performance of the Agricultural Sector under Structural Adjustment', in E. Friis-Hansen (ed.), *Agricultural Policy in Africa after Adjustment*, Copenhagen: Centre for Development Research Policy Paper, September, pp. 24–34.

Goetz, A. M. and R. Jenkins (1999), 'Creating a Framework for Reducing Poverty: Institutional and process issues in national poverty policy—Uganda Country Report', Draft report of a DFID and Sida funded study 'Creating a Framework for Reducing Poverty: Institutional and Process Issues in National Poverty Policy in Selected African Countries' for the Poverty and Social Policy Working Group (PSPWG) of the Special

Programme of Assistance, April, http://www.ids.ac.uk/ids/pvty/uganda.pdf

Government of the Republic of Uganda GoU (1998), *Towards a Sector Wide Approach: Developing a Framework for the Modernization of Agriculture in Uganda*, Statement to the December 1998 Consultative Group Meeting.

———— (2000), *Plan for the Modernization of Agriculture: Eradicating Poverty in Uganda*, Kampala: Ministry of Agriculture, Animal Industry and Fisheries, Entebbe: Ministry of Finance, Planning, and Economic Development.

Greeley, M. and R. Jenkins (1999), 'Mainstreaming the Poverty-Reduction Agenda: An Analysis of Institutional Mechanisms to Support Pro-poor Policy Making and Implementation in Six African Countries', Paper prepared for the meeting of the SPA Working Group on Poverty and Social Policy, 19–21 October, Paris, http://www.ids.ac.uk/ids/pvty/paris.pdf

Hannig, A. and Wisniwski (1999), *Mobilizing the Savings of the Poor: Experience from Seven Deposit-taking Institutions*, Eschborn: GTZ, January (mimeo).

Hansohm, D. and W. Thomas (2001), 'Development: An Illusion for Africa?' D+C (Development and Cooperation) No. 3/2001, Frankfurt-am-Main: DSE, pp. 13–15.

Hilhorst, T. (2000), 'Women's Land Rights: Current developments in Sub-Saharan Africa', in C. Toulmin and J. Quan (eds), *Evolving Land Rights, Policy and Tenure in Africa*, London: DFID/IIED/NRI, pp. 181–96.

Holmgren, T., L. Kasekende, M. Atingi-Ego, and D. Ddamulira (1999), *Aid and Reform in Uganda: Country Case Study, Aid effectiveness research*, Washington, DC: World Bank, http://www.worldbank.org/research/aid/africa/uganda.pdf

Kappel, R. (2001), 'The End of the Great Illusion', D+C (Development and Cooperation) No. 2/2001, Frankfurt-am-Main: DSE.

Kayani, R. and A. Dymond (1997), *Options for Rural Telecommunications Development*, World Bank Technical Paper No. 359, Washington, DC: World Bank.

Kidd, A. D. and B. Schrimpf (2000), 'Bees and Bee-keeping in Africa', in R. M. Blench and K. C. MacDonald (eds), *The Origins and Development of African Livestock: Archaeology, genetics, linguistics and ethnography*, London: UCL Press, pp. 503–26.

Klugman, J., B. Neyapti, and F. Stewart (1999), *Conflict and Growth in Africa*, Vol. 2: *Kenya, Tanzania and Uganda*, Paris: Development Centre, OECD.

Lecup, I. and K. Nicholson (2000), *Community-based Tree and Forest Product Enterprises: Market analysis and development*, Rome: FAO.

Longley, K. (2001a), 'Farming Systems of the Acholi: A comparison between the status now and before the start of conflict', London: Overseas Development Institute (draft report).

———— (2001b), 'Beyond Seeds and Tools: Opportunities and challenges for alternative interventions in protracted emergencies', *Humanitarian Exchange*, No. 18, pp. 6–9, London: Humanitarian Practice Network (HPN), Overseas Development Institute.

Luckham, R., I. Ahmed, R. Muggah, and S. White (2001), *Conflict and Poverty in Sub-Saharan Africa: An assessment of the issues and evidence*, IDS Working Paper 128, Brighton: Institute of Development Studies at the University of Sussex.

Lyon, F. (2000), 'Trust, Networks and Norms: The creation of social capital in agricultural economies in Ghana', *World Development*, Vol. 28, No. 4, pp. 663–81.

MAAIF (2000a), *The National Agricultural Advisory Services Programme: Master Document of the NAADS Task Force and Joint Donor Group*, Entebbe: Ministry of Agriculture, Animal Industry and Fisheries, October.

———— (2000b), *Sector Plan to address HIV/AIDS*, Entebbe: Ministry of Agriculture, Animal Industry and Fisheries.

Mackinnon, J. and R. Reinikka (2000), *Lessons from Uganda on Strategies to Fight Poverty*, Policy Research Working Paper No. 2440, Washington, DC: World Bank, *http://econ.worldbank.org/docs/1195.pdf*

Matovu, J. M. and F. Stewart (2000), 'The Social and Economic Costs of Conflict: Uganda, a case study', in F. Stewart, and V. Fitzgerald (eds), *War and Underdevelopment: Case Studies in Country of Conflict*, Vol. 2, Oxford: Oxford University Press.

McGee, R. (2000), 'Analysis of Participatory Poverty Assessment (PPA) and household survey findings on poverty trends in Uganda', *http://nt1.ids.ac.uk/eldis/fulltext/mcgee.pdf*

MFPED (2001), *Uganda Poverty Reduction Strategy Paper Progress Report 2001: Summary of Poverty Status Report*, Kampala: Ministry of Finance, Planning and Economic Development, 8 February.

———— (2000a), *Poverty Reduction Strategy Paper: Uganda's Poverty Eradication Action Plan Summary and Main Objectives*, Kampala: Ministry of Finance, Planning and Economic Development, 24 March 2000.

———— (2000b), *Uganda Participatory Poverty Assessment Report: Learning from the Poor*, Kampala: Ministry of Finance, Planning and Economic Development, June.

———— (2000c), 'Medium-term Competitive Strategy for the Private Sector (2000–2005)', Kampala: Ministry of Finance, Planning and Economic Development.

———— (1997), *Report on the Economics of Crops and Livestock Production*, Kampala: Agricultural Policy Committee, Ministry of Finance, Planning and Economic Development.

MGLSD and UBOS (2000), *Women and Men in Uganda—Facts and Figures*, Sectoral Series: Decision-Making, Kampala: Ministry of Gender, Labour and Social Development, and Entebbe: Uganda Bureau of Statistics.

Moore, M. and J. Putzel (1999), 'Politics and Poverty: A background paper' for the *World Development Report 2000/1*, Brighton: Institute of Development Studies at the University of Sussex.

Morrissey, O. (2000), *Case Studies of the Poverty Experience in Economies Undergoing Economic Adjustment*, Consultancy report as input for the DFID White Paper on 'Eliminating World Poverty: Making Globalization Work for the Poor', Nottingham: CREDIT, University of Nottingham; London: ODI; and London: DFID Economists' Resource Centre, *http://62.189.42.51/BackgroundWord/CaseStudiesOliverMorrissey.doc*

Nafziger, E. W. and J. Auvinen (1997), *War, Hunger and Displacement: An econometric investigation into the sources of humanitarian emergencies*, WIDER Working Paper No. 142, Helsinki: World Institute for Development Economics Research.

Nakileza, B. and E. N. B. Nsubuga (eds) (1999), *Rethinking Natural Resource Degradation in Semi-Arid Sub-Saharan Africa: A review of soil and water conservation research and practice in Uganda, with particular emphasis on the semi-arid areas*, Kampala: Soil and Water Conservation Society of Uganda (SWCSU), Makerere University; London: Overseas Development Institute.

Neuchâtel Group (2000), *Guide for Monitoring, Evaluation and Joint Analyses of Pluralistic Extension Support*, Eschborn: GTZ; Stockholm: Sida and Bern: SDC.

——— (1999), *Common Framework on Agricultural Extension*, Paris: Ministère des Affaires étrangères.

NTCA (2000), *Initial Lessons Learned About Private Sector Participation in Telecentre Development: A Guide for Policy Makers in Developing Appropriate Regulatory Frameworks*, Arlington, VA: National Telephone Cooperative Association, *http://www.ntca.org/intl/FINAL.pdf*

Nyanzi, T., C. Healy, and P. Bevan (1998), *Exploring the Dynamics of Poverty in Africa: A regional case study from Uganda*, Report to the UK Department for International Development (ESCOR Research project R7165), Department of Economics and International Development, University of Bath.

O'Farrell, C., P. Norrish, and A. Scott (1999), *Information and Communication Technologies (ICTs) for Sustainable Livelihoods: Preliminary Study April–November 1999*, Reading: AERDD; London: ITDG, *http://www.rdg.ac.uk/AERDD/AERDD/Csds.htm*

OCHA (2001), *Humanitarian Update—Uganda*, monthly newsletter produced by the UN Office for the Co-ordination of Humanitarian Affairs, Kampala, *http://www.db.idpproject.org/Sites/idpSurvey.nsf/wViewCountries/0FEFFAFF14EF7217C125686B00368215?OpenDocument*

Okoth-Ogendo, H. W. O. (2000), 'Legislative Approaches to Customary Tenure and Tenure Reform in East Africa', in C. Toulmin and J. Quan (eds), *Evolving Land Rights, Policy and Tenure in Africa*, London: DFID/ IIED/NRI, pp. 123–34.

Oxfam (2001), *Conflict's Children: The human cost of small arms in Kitgum and Kotido, Uganda: A case study*, Oxford: Oxfam, *http://www.oxfam.org. uk/policy/papers/Ugandaarms.pdf*

Palmer, R. (2000), 'Land Policy in Africa: Lessons from Recent Policy and Implementation Processes', in: C. Toulmin and J. Quan (eds), *Evolving Land Rights, Policy and Tenure in Africa*, DFID/IIED/NRI, pp. 267– 88.

Prakash, S. (2000), *Uganda: Information Technology and Rural Development— The Nakaseke Multi-Purpose Telecenter*, IK Notes No. 27. Washington, DC: World Bank, *http://www.worldbank.org/afr/ik/iknt27.pdf*

Robbins, P. (1999), *Review of the Impact of Globalization on the Agricultural Sectors and Rural Communities of ACP Countries*: A study commissioned by the Technical Centre for Agricultural and Rural Cooperation (CTA), Study Report 4–1–06–211–9, London: CMIS, *http://www.agricta.org/ pubs/globalreview/globalreview.pdf*

Robbins, P. and S. Ferris (2000), *Design of a Market Information System for Small-scale Producers and Traders in Three Districts of Uganda*, CTA No. 8019, Kampala: IITA/Foodnet Project; London: Commodity Marketing Information Services (CMIS), *http://www.agricta.org/pubs*

Rutherford, S. (1999), *Savings and the Poor: The methods, use and impact of savings by the poor of East Africa*, Kampala: MicroSave-Africa East African Savings Study.

——— (1996), *A Critical Typology of Financial Services for the Poor*, London: Action Aid and Oxfam.

Scialabba, N. (2000), 'Factors Influencing Organic Agriculture Policies with a Focus On Developing Countries', Paper presented at the IFOAM 2000 Scientific Conference, Basel, Switzerland, 28–31 August.

Sebstad, J. and M. Cohen (2000), 'Microfinance, Risk Management, and Poverty, Assessing the Impact of Microenterprise Services (AIMS)', Study submitted to the Office of Microenterprise Development, Washington, DC: USAID, *http://www.mip.org/PDFS/AIMS/WDR per cent20 REPORT.pdf*

Stewart, F. (2000), *Crisis Prevention: Tackling Horizontal Inequalities*, QEH Working Paper Series No. 33, Oxford: Queen Elizabeth House, University of Oxford, *http://www2.qeh.ox.ac.uk/pdf/qehwp/qehwps33. pdf*

——— (1998), *The Root Causes of Conflict: Some Conclusions*, QEH Working Paper Series No. 16, Oxford: Queen Elizabeth House, University of Oxford, *http://www2.qeh.ox.ac.uk/pdf/qehwp/qehwps16.pdf*

Stewart, F. and M. O'Sullivan (1998), *Democracy, Conflict and Development:*

*Three Cases*, QEH Working Paper No. 15, Oxford: Queen Elizabeth House, University of Oxford, *http://www2.qeh.ox.ac.uk/pdf/qehwp/qehwps15.pdf*

Tangri, R. and A. Mwenda (2001), 'Corruption and Cronyism in Uganda's Privatization in the 1990s', *African Affairs*, Vol. 100, pp. 117–33.

Tele Commons Development Group (2000), 'Rural Access to Information and Communication Technologies: The Challenge for Africa', Paper prepared for the African Connection Secretariat with support from the Information for Development Program (*info*Dev), Washington, DC: World Bank, *http://www.infodev.org/projects/afcon/afconreport.pdf*

Therkildsen, O. (2001), *Efficiency, Accountability and Implementation of Public Sector Reform in East and Southern Africa*, Democracy, Governance and Human Rights Programme Paper No. 3, Geneva: United Nations Research Institute for Social Development, February.

Topouzis, D. (2000), 'The Impact of HIV on Agriculture and Rural Development: Implications for training institutions', in *Human Resources in Agricultural and Rural Development 2000*, Rome: FAO, pp. 93–103.

Topouzis, D. and G. Hemrich (1996), *The Socio-economic Impact of HIV/AIDS on Rural Families in Uganda*, UNDP Discussion Paper No. 6, New York: UNDP.

Topouzis, D. and J. du Guerny (1999), *Sustainable Agricultural/Rural Development and Vulnerability to the AIDS Epidemic*, Geneva: UNAIDS, Rome: FAO.

Transparency International (2000), *Corruption Perceptions Index 2000*, *http://www.transparency.de/documents/cpi/2000/cpi2000.html*

Tsikata, Y. (2000), *Globalization, Poverty and Inequality in Sub-Saharan Africa: A political economy appraisal*, Paris: OECD Development Centre, *http://www.oecd.org/dev/ENGLISH/pagelisteE/Poverty-Ineq/Documents/Povertyincome_subsahafrica.pdf*

Toulmin, C. (2000), 'Decentralization and Land Tenure', in C. Toulmin and J. Quan (eds), *Evolving Land Rights, Policy and Tenure in Africa*, London: DFID/IIED/NRI, pp. 267–88.

UBOS (2001), *Uganda National Household Survey 1999/2000*, Reports on the Community Survey and on the Socio-Economic, Entebbe: Uganda Bureau of Statistics, *http://www.ubos.org/nhsrepo.html*

UNAIDS (1996), 'HIV/AIDS Epidemiology in Sub-Saharan Africa', Fact Sheet No. 1, New York: UNAIDS.

World Bank (2002), 2002 World Development Indicators CD Rom, Washington, DC: World Bank.

——— (1996), *Performance and Perceptions of Health and Agricultural Services in Uganda*, A Report Based on the Findings of the Baseline Service Delivery Survey, Washington, DC: Economic Development Institute and CIET International.

Wright, G. (1999), *A Critical Review of Savings Services in Africa and*

*Elsewhere*, Kampala: Centre for MicroFinance, *http://www.undp.org/ sum/MicroSave/ftp_downloads/wright.pdf*

Wright, G. A. N., D. Kasente, G. Ssemogerere, and L. Mutesasira (1999), *Vulnerability, Risks, Assets and Empowerment—The Impact of Microfinance on Poverty Alleviation. MicroSave-Africa and Uganda Women's Finance Trust*, Kampala: MicroSave-Africa and UWFT Final Report, *http:// www.undp.org/sum/MicroSave/ftp_downloads/UWFTstudyFinal.pdf*

# 5

# Extension, Poverty, and Vulnerability in Nicaragua

*Ian Christoplos*

## THE COUNTRY CONTEXT

### BASIC INDICATORS

With an annual per capita income of US$ 430 (1999), Nicaragua is the second poorest country in Latin America. Internationally, Nicaragua is in the highest 20 per cent in terms of inequity of income (Government of Nicaragua 2000). Whereas life expectancy is near the average for Latin America, most other indicators are far lower, particularly in rural areas (UNDP 2000). Fertility rates are double the average for Latin America, and adolescent fertility is the highest in the region (World Bank 2000). Poverty is closely correlated to youth. Twenty per cent of children under five are chronically undernourished or stunted (see also Table 5.1). Agricultural productivity, as measured by production per unit of land area, is considerably lower than that in other Central American countries.

### RURAL POVERTY

Poverty is highly concentrated in the countryside, particularly in the areas that were most affected by the conflict of the 1980s, and where there is relatively limited commercialization (see Map 5.1).

There was a gradual collapse of the economy during 1978–94, before which Nicaragua had a relatively strong, though highly inequitable economy, particularly in agriculture. Relative poverty rates declined slightly during the late 1990s, primarily in urban areas, but absolute rates have increased. Despite a modest recovery during the

TABLE 5.1
Nicaragua, Basic Indicators

| Series | Value | Year |
|---|---|---|
| Cereal yield (kg per hectare) | 1842 | 2000 |
| Land use, arable land (hectares per person) | 0.5 | 1999 |
| Land use, irrigated land (% of cropland) | 3.2 | 1999 |
| Agriculture, value added (% of GDP) | 32 | 2000 |
| GNI per capita, Atlas method (current US$) | 400 | 2000 |
| Population, total | 5,071,000 | 2000 |
| Rural population (% of total population) | 35 | 2000 |
| Malnutrition prevalence, height for age (% of children under 5) | 25 | 1998 |
| Malnutrition prevalence, weight for age (% of children under 5) | 12 | 1998 |
| Low birthweight babies (% of births) | 8 | 1995 |
| Poverty headcount, national (% of population) | 50 | 1993 |
| Poverty headcount, rural (% of population) | 76 | 1993 |
| GINI index | 60 | 1998 |
| Mortality rate, infant (per 1000 live births) | 33 | 2000 |
| School enrolment, primary (% net) | 77 | 1997 |
| Surface area (sq. km) | 130,000 | 2000 |
| Roads, total network (km) | 19,032 | 2000 |

*Source:* World Bank (2002).

late 1990s, per capita gross domestic product (GDP) is approximately half that of the 1960s and 1970s (World Bank 2000). External debt amounts to 600 per cent of exports and is three times the annual GDP (UNDP 2000). In order to qualify for the highly indebted poor country (HIPC) Initiative, the size of the public sector has been scaled back considerably, from a high of nearly 250,000 employees during the Sandinista era to fewer than 80,000 today. In compliance with the HIPC conditions, Nicaragua has developed a poverty reduction strategy paper (PRSP). In addition to macroeconomic and structural factors, natural disasters and complex political emergencies are central aspects of the vulnerability context of Nicaraguan development (see Box 5.1).

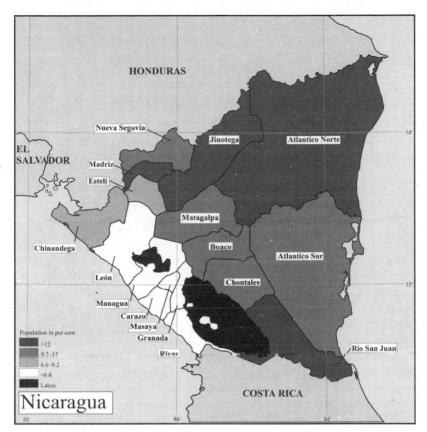

1994 MAGELLAN GeographixSM Santa Barbara, CA (800) 929–4627

MAP 5.1: Spatial distribution of poverty in Nicaragua by department

*Source:* ERRP (2000) *Estrategia Reforzada de Reducción de la Pobreza*, Secretaría Técnica de la Presidencia de la República (SETEC), Gobierno de Nicaragua, Managua, Agosto 2000, and accessed at:
*http://www.fao.org/geonetwork/images/largegifs/3744.gif*

The sum result of these events is a dynamic of changing vulnerability. Hurricane Mitch did not affect the overall poverty profile in Nicaragua (World Bank 2000), though it did have a profound impact on the livelihood strategies of the poor. Development has, therefore, not been a linear process. Shocks to livelihoods and to the national economy and public expenditure are regular occurrences. In the northwestern part of the country, there is a 25 per cent chance of major

```
                            BOX 5.1
                       Chronology of Events
        1972          Managua  earthquake
        1979          Sandinista revolution
        1983–90       Civil war and United States embargo
        1988          Hurricane Joan
        1992          Tidal wave
        1992, 1994    Volcanic eruptions
        1996–8        El Niño drought
        1997          Hurricane Mitch
        2000–present  Collapse in coffee prices
```

agricultural losses due to drought in any given year (World Food Programme (WFP) 2001). Resilience is perhaps, in many ways, a more sensible objective than stable growth. Studies have shown that the poor perceive the increased risk of their current situation (with a market economy, uncertain safety nets, etc.) as being a major aspect of their poverty. They react by adopting risk-averse production strategies (World Bank 2000).

Nicaragua displays a curious combination of areas where a seemingly 'normal' process of economic development is underway, with areas within a relatively short distance from the capital where insecurity and violence continue. The peace accords included promises of land and rural services that have in many cases not been implemented. Consolidation of the peace process is still not complete (Ardón 1999), due not least to the legacy of debt and decline in social capital inherited from the war years (FitzGerald and Grigsby 2001). Rural public services are very weak, because of Nicaragua's extremely limited public finances, and the conditions subsequently followed as part of qualifying for debt relief within the HIPC process. Corruption levels are high, and the donor community has followed an exceptionally firm, frank, and openly critical dialogue with the government on the issue of transparency.

POLICIES TOWARDS AGRICULTURAL AND
RURAL DEVELOPMENT

Given its extreme indebtedness and geo-political position as a small country with very close links with the United States, Nicaragua has very little capacity to withstand pressures of globalization. Furthermore, its current neo-liberal government has embraced open markets as a solution for economic development, and with that for poverty alleviation. Globalization has two basic impacts on markets that affect the poor in Nicaragua.

First, is access to export markets. Indications of success in taking advantage of export opportunities are mixed. Growth has been good since the mid-1990s, but this can be seen to a large extent as a recovery from near-collapse at the end of the 1980s. Lack of infrastructure, weak entrepreneurialism, poorly functioning credit markets, fragmented institutions, and poor governance constitute major obstacles to even the wealthier actors in the agricultural economy drawing benefits from globalization. Most agricultural service providers in Nicaragua are pessimistic that poor producers will succeed in significantly accessing international markets.

The second, and perhaps more relevant, question about the impact of globalization is whether poor producers will be able to retain a domestic market. Regional imports are increasingly dominating the domestic market. Despite a relative abundance of land and labour, Nicaragua lags far behind its Central American neighbours in agricultural productivity. Traditionally, Nicaragua was able to compete largely by expanding production areas in the 'agricultural frontier' of the former rain forest, without increasing productivity. With the destruction of the forest, this is no longer a significant option. Nicaragua must now catch up with its more populous neighbours by adopting more intensive production systems. So far, however, Nicaraguan labourers have in many cases been able to better enhance their livelihoods as migrants on better-capitalized, more market-oriented, and infrastructurally accessible Costa Rican farms than they can at home. Simple assumptions that cheap land and labour automatically constitute a comparative advantage are not valid in the Nicaraguan case.

Although expansion of the agricultural frontier is no longer a major option, Nicaragua is not overpopulated. Favourable agro-ecological conditions and relatively abundant land present opportunities to increase production through intensification. Demographically, an 'escape' or exit from agriculture would not seem essential, even though most analyses of poverty show that this is the most attractive option.

In some areas, Nicaraguan producers have been able to compete. Milk and cheese exports to El Salvador and Honduras have done well.

Beans are a traditional product for which demand remains strong, even in urban markets (unlike maize, which is losing ground to other grains). Vegetable production is growing, despite regional imports. Coffee production has recovered, even though the current low prices mean that significant further investment is now largely on hold.

Extension and agricultural priorities must be seen in the perspective of economic trends and poor people's livelihoods, both of which point towards good exits from farming being as important for rural development as improvement in farming itself. Fifty per cent of rural income is derived from non-agricultural activities, and education levels can be directly correlated to the ability of rural households to diversify out of agriculture (Government of Nicaragua 2000). This is in line with trends elsewhere in Latin America (Berdegué *et al.* 2000). Non-farm incomes are derived mainly from services in relatively accessible areas (Corral and Reardon 2001).

Migration is particularly important. Close to half of farm households have at least one family member permanently working away from the farm, of which a large proportion are outside Nicaragua (WFP 2001). Given the importance of migration and remittances for the poor, an awareness has emerged that migration has not just been a drain on rural communities, but also a major factor in keeping rural communities alive and bringing in much needed capital.

Nicaragua is an example of a country struggling with post-conflict and post-natural disaster issues. This involves coping with the massive destruction, extremely high debt, and economic collapse stemming from both forms of disaster. It is also impacted by large aid flows and an entrenched focus on 'projects' as the motor for rural development. Simplistic polemics based on dichotomies between 'dependency' and 'sustainability' fail to provide a basis for understanding the complex landscape of rural development in Nicaragua today. In order to effectively analyse Nicaraguan rural development, it is essential that 'abnormal' events (such as disasters) and structures (such as the active donor engagement in the national policy discourse) be accepted as part of the context of policy formation during the coming decade.

Several policy frameworks are of major relevance in relating poverty and vulnerability to extension. These strategies share a broad acceptance that Nicaragua must face globalization head-on. Continued structural reform and open markets are inevitable. Strategies alternate, however, between assuming that explicit measures are necessary to ensure inclusive development, and assumptions that growth alone will eradicate poverty. The policy formation process in Nicaragua has been profoundly influenced by the experience of Hurricane Mitch, and the relatively massive aid flows that followed. The context before Mitch was one of polarization between a neo-liberal governing regime and an opposition of the populist left. This state of affairs has shifted to a more complex set of forces involving donors as active policy advocates, a more united and stronger set of civil society institutions, and a government pressured more towards populism in the face of national elections.

Nicaragua's 'Strengthened Poverty Reduction Strategy' is the main policy initiative that takes a livelihoods approach to analysing how the poor employ their assets. The strategy emphasizes that poverty is primarily a rural phenomenon, but that even in rural areas, the primary way to escape from poverty is to move away from agriculture, particularly from subsistence agriculture. Areas with least poverty have access to labour markets. Those with the highest levels of basic cereal production have the highest levels of malnutrition.

The Ministry of agriculture, livestock, and forestry (MAGFOR) takes a very different perspective. It has three basic goals: (i) the rationalization of production, (ii) institutional modernization, and (iii) food security. Productivity increase is the central focus. The strategy is supportive of those farms with the capacity to take advantage of market opportunities and to make major productivity leaps, and is thus most viable in those areas of the country that have relatively good access to markets. Before Hurricane Mitch, food security had received very little attention. After considerable criticism of failure to address extreme poverty and vulnerability, the government is now placing greater emphasis on food security. The current policy combines a focus on food crops with presumptions that productivity increase will solve food insecurity. There is a notable tendency to divorce food security objectives from the context of poverty. People are assumed to be food insecure due to their low production levels rather than due to the poverty that limits their capabilities, not only to produce food, but also to gain entitlements as labourers. Donors, non-governmental organizations (NGOs), and United Nations (UN) agencies, however, are actively engaged with the government in promoting a deeper understanding of food security and vulnerability.

The link between inappropriate agricultural and natural resource management practices and a heightened risk of natural disasters is central to the inclusion of vulnerability in rural development policy. This is reflected in the agreement among the governments, civil society, and donors on principles for the 'transformation' of Central America after Hurricane Mitch. The 'Stockholm Declaration' high-lighted the problems of weak governance, political polarization, and the lack of co-ordination capacity in the massive reconstruction effort, while positioning poverty and environmental risk within the reha-bilitation and development agenda.

## AGRICULTURAL EXTENSION: BACKGROUND AND STATUS

Until recently, the current government pursued neo-liberal agricul-tural policies with genuine commitment. A minimal role for the state in service provision was accepted and, despite difficulties in rational-izing staffing, extension was expected to be a showcase for the reform effort, with services being increasingly contracted out to the private sector. Consideration of 'public goods' issues was a major feature in the design of new structures. In the run up to the 2001 elections, these issues began receiving less attention. Proposals were made to establish a national extension structure based on broad coverage and very intensive extension agent to farmer contacts in order to invigorate agricultural development. This shift has emerged from the broader political context. The failures of the government to mobilize a strong response to Hurricane Mitch were rooted in neo-liberal policies that reduced public service capacity. Political pressures to shift to more populist policies are growing, and extension agents working face-to-face with farmers were seen as an effective way to demonstrate government commitment.

The ebb and flow of policy reform has been influenced by three narratives. The first (and formerly dominant) was a set of neo-liberal concepts based on a minimal role for government agencies in imple-menting programmes, paired with a broad faith in economic growth as the driving force both supporting and deriving from agricultural development. As elections drew near, this gave way to an alternative narrative that places production growth at the centre of strategic thinking. A pragmatic and simpler drive emerged to get services to farmers. Questions of who and how (and the longer-term sustainability of the 'whos' and 'hows') were put on the back burner in the interest of showing results and stimulating a rapid transformation. The third narrative is that of vulnerability reduction and poverty alleviation.

This agenda, promoted primarily by the donor community and civil society, acknowledges that neither economic nor productivity growth will automatically address the deplorable situation of the poor.

The institutional landscape in Nicaragua contains a confusing and seemingly paradoxical mix of policies, structures, and priorities (Box 5.2). Non-governmental organizations that often trace their roots to leftist initiatives are actively promoting a modest role for the government and stronger market orientation. State bureaucracies, although led by the neo-liberal government, have been slow to adopt a market focus and have plans to expand their roles. Furthermore, Nicaragua is a land of projects. Government capacity to use policy as a tool to co-ordinate the mass of projects that together make up the thrust of Nicaraguan rural development initiatives has been limited. 'Projectization' has a profound impact on the nature of institutions offering extension services. Agencies expect to be judged by donors on their potential capacity to undertake different extension tasks, rather than 'correct' service provision slots for state, private sector, and civil society institutions.

---

BOX 5.2

Where Policies Fail to Influence Praxis—The Nicaragua Case

Nicaragua is a land of projects, but government capacity to use policy as a tool to co-ordinate the mass of projects has been limited. The reasons for this are:

- a political process is entrenched, wherein the tool of patronage via donor funded projects tends to overshadow policy vision;

- there is a generally genuine (though perhaps fading) commitment to a neo-liberal ideology that sees the role of the state as very limited;

- the state has very limited capacity to mobilize its own resources (due to debt service and HIPC restrictions);

- large and unpredictable aid flows tend to overshadow modest state resources;

- profound donor concerns exist regarding corruption and lack of transparency, which in turn encourage bypass solutions;

- street-level bureaucracies frequently lack awareness of and interest in official government policies;

- government policies lack legitimacy in the field due to a widespread perception that they are steered by personal whims and interests of current (and highly interchangeable) ministers and, therefore, do not represent a consistent framework for action.

All these factors have meant that *de facto* development policy formation is extremely fragmented by projects. The impacts of the high level of projectization of rural development on policy narratives are:

- a strong 'supply side' bias exists, where concern over how to fund and implement a given agency's (or donor's) preferred solution takes precedence over a given action's relevance to policy objectives or to the livelihood and asset investment strategies of the poor;

- demand-pull mechanisms are overshadowed by pipeline pressures and paternalism;

- there is a lack of continuity in service provision and in relationships between service providers and their clients, contributing further to a lack of demand mechanisms;

- little attention is given to defining roles of different institutions based on public goods;

- an extreme fragmentation of services exists, where rural people have little control over the continuity, quality and priorities of service provision;

- pluralism in the provision of services has not resulted in pluralism in options for producers (with the notable exception of certain easily accessible areas and geographical foci of the 'CNN effect' after Hurricane Mitch); and

- there are strong tendencies toward a 'contract culture' among service provision agencies.

Decentralization of responsibilities for natural resource management and the projectization of rural development have created a potential for greater subsidiarity in extension and agricultural development. As yet, there are relatively few examples of this potential being acted upon either by local governments, line ministry agencies, or the various actors managing rural development projects.

This is due to several factors:

1. local government has limited institutional and financial capacity in rural development;

2. local political priorities focus on urban development and infrastructure because politicians and their constituencies assume that this is the role of municipal government;

3. there are virtually no lines of accountability from public sector agricultural institutions to local government;

4. cynicism and pessimism prevail among donors and NGOs on the potential for strengthening the local government's role outside of urban areas; and

5. paternalism and prevalence of donor-driven agendas hinder attempts to strengthen local government, leading to a lack of genuine ownership.

Extension structures have followed the overall national trends of expansion and contraction of the public sector. Nicaragua had a large public sector (24 per cent of the work force) in 1990, which was reduced to 5.3 per cent by 1998. The World Bank supported extension programme in Nicaragua has taken a lead in introducing user charges for extension services and in contracting out service provision to private firms. Producer organizations and NGOs usually have very negative preconceived views of user-charges, though they have little experience in their use.

Non-governmental organizations are involved in farmer-to-farmer approaches promoting watershed management, sloping agricultural land technologies, home gardens, and alternatives to slash and burn agriculture through both concrete extension projects and advocacy. Such projects have succeeded in establishing a certain level of national debate on alternatives to conventional agriculture. Some doubts exist, however, about the longer-term financial viability of these types of extension programmes. The agricultural technologies themselves may (perhaps) be profitable. Critics, however, point out that the rhetoric about farmers helping one another may hide a considerable level of donor-funded investment in extension staff and logistics. Before these approaches become more definite mainstream alternatives to conventional extension programmes they will first need to be subjected to the same scrutiny as other initiatives. That said, the cost of farmer-to-farmer approaches could be justified, based on the reduced levels of environmental destruction.

ROLE OF EXTENSION

A large extension structure was first developed during the 1970s with support from the United States Agency for International Development (USAID). This was followed by a broader extension-led structure during the Sandinista years, wherein extension agents became development agents, with a broad range of rural development roles. All of this collapsed with the economic crisis at the end of the 1980s and early 1990s.

During both these periods, the role of extension was primarily to support large-scale farmers. During the Somoza era, economic development was driven by wealthy enterprises run by the Somoza family and its associates. The Sandinistas primarily promoted large-scale, state and co-operative run agro-industrial enterprise, excluding small producers (Maldidier and Marchetti 1996). This created a strong distrust of the Sandinistas (which still exists today) among many small-scale farmers, particularly in northern Nicaragua. In general, technology transfer has been characterized by an elite, high external input and capital-intensive bias, and has essentially subsidized the production of better-off farmers (Báez and Baumeister 1997).

The National Institute for Agricultural Technology (INTA) was founded in 1993, with support from the World Bank and the Swiss Agency for Development and Co-operation (SDC). Its main roles are research (primarily validation trials) and extension services, but it is also engaged in seed multiplication. After Hurricane Mitch, INTA started managing food-for-work projects. It was created as a semi-autonomous institution, as a reaction against the experience of the Sandinista years when the extension service became a powerful but overburdened tool for the broad implementation of rural development policy. It was to be managed outside of line ministry structures to ensure efficient implementation of policy, and to avoid politicization. In 1998, INTA was administratively placed under the MAGFOR, but with its principle of autonomy largely intact. Semi-autonomous research and extension institutions have had a long history in Latin America (Nogueira 1990).

With its flow of donor funding, INTA is relatively well-financed. Its 150 field staff and 125 additional private sector contracted staff are mobile and relatively well-trained, though some question their capacity for innovation and efficiency. The World Bank support to INTA has, in recent years, taken a lead in introducing user-charges for extension services and in contracting out service provision to private firms. Farmers are charged a set proportion of the costs of direct service provision in areas deemed to have high potential. This has been seen as a model for introducing cost-recovery in other countries (Dinar and Keynan 1998). More recently, this model has been acknowledged to have had mixed results. Willingness to pay for services has varied, but the experience has been mainly positive.

The geographic coverage of INTA is, however, limited. It has little coverage in poorer areas and virtually none on the Atlantic coast, the main target for food security and poverty alleviation initiatives. The private firms providing services are said to emphasize the more accessible clients even more (Dinar and Keynan 1998). Although this is acknowledged as a problem, there is no clear strategy of how to address the spatial nature of poverty. There seems to be an implicit assumption that the poorest areas will be served by donor projects.

NGOs and producer organizations frequently have very negative preconceived views of direct user charges, though they have little experience of the practice. Despite this mistrust of cost-recovery in extension *per se*, NGOs are even more involved than INTA in market integration through improved quality control, certification, and processing. As producers gradually take on the responsibility for such schemes, NGOs are starting to accept that producers will need to cover the costs of advisory services within broader packages.

Neither the government nor NGOs expect that service charges or contracting out will be a viable way to support isolated farmers engaged in subsistence production and home gardens. Many expect that environmental protection in particular (primarily in the form of watershed management programmes) will require further subsidies for the foreseeable future, both for recurrent costs and even in the form of food or cash for work.

The vision of the World Bank is to encourage INTA to develop into a market-oriented 'think tank' supporting a multiplicity of private sector service providers, and using a combination of public and private finance. The assumption is that as the market for contracted service provision expands, so will the supply and quality of private sector services, some of which will be charged for and others not. The INTA will thereby gradually withdraw from direct service provision. It is acknowledged, however, that this transformational process within INTA will be difficult. Current staffing and organizational culture are focused on 'doing extension', and INTA is, for the time being, ill-equipped for this new role.

In the meantime, INTA has sometimes been tempted to enter the competition for managing projects, effectively drawing it in an opposite direction. In particular, it has become involved in contracting for the management and support of food-for-work schemes. This is seen by INTA and by the WFP as a positive way to increase links with poorer groups. Being 'contracted-in' by such aid programmes may also have implications for INTA's financial stability as the sources of revenue are diversified. There may, however, be a downside to INTA's engagement in project contracting. As field staff take on potentially diverse and short-term activities, their clients may receive less coherent and regular contacts with extensionists. If INTA begins marketing their services upwards to funding agencies, their downward accountability to farmers could suffer accordingly.

There are two basic extension approaches of INTA. Services to 'favourable areas' are based on direct advisory services with a degree of cost recovery. Intensification and diversification are accorded priority in the favourable areas. 'Less-favoured areas' are addressed through a programme of mass technical assistance (*assistencia tecnica massiva*, ATM). This modality was originally structured on the use of media, farmer fairs, etc. The results were deemed unsatisfactory, and this component is currently under review. In practice, ATM focuses on lead or model farmers, with subsistence, nutrition (home gardens), and environmental protection as major components. Low-risk maize and sorghum varieties are promoted. Gender issues are increasingly receiving explicit attention.

In practice, extensionists are given leeway to respond to producer demands, regardless of whether they live in a favourable or less-favoured areas. This often results in more emphasis on cereals, even in areas classified (based largely on rainfall and soil quality data) as having higher potential. Choice of technologies may also relate to the relative strengths and weaknesses of INTA's technological portfolio, since it is known to have more to offer in relation to basic grains than cattle or coffee, for example. There are also some indications that farmers' demands tend to correspond with a desire to access whatever free inputs INTA has available at a given time. In addition, farmers report that they value INTA's services largely as an avenue for preferential access to credit (Barandun 2001).

Greater emphasis is now being placed on collaboration with other institutions (NGOs, IDR, local government, etc.). Partnerships focus primarily on natural resource management and soil conservation (often including food-for-work), inspired by the experience of Hurricane Mitch. This focus has the potential to direct INTA away from wealthier farmers and toward closer collaboration with NGOs. The INTA is also increasing its collaboration with NGOs in post-harvest technologies.

Clear analyses of public goods were given attention in the original design and planning of the second phase of World Bank support to INTA. Efforts were made to specify which services could be provided on a commercial basis, and which would need to be provided for free. In the first phase, there was, however, a general impression among many observers that those aspects defined as public goods perhaps received less attention, leading to the poor performance of the ATM modality. Many of the deficiencies in balancing efforts to satisfy demand for both public and private goods have been addressed in the planning of the next phase.

The MAGFOR itself (excluding INTA structures), in principle limits its role to normative inputs, information flow, sanitary protection, and certification. Its services are entirely oriented towards regulatory efforts and public goods. Its greatest successes in service provision have been in parasite control, particularly the eradication of the Barrenador worm (MAGFOR 2001). It has a structure of provincial delegations that are frequently perceived of as weak and politicized. Steps are being taken to reduce politicization and improve the stature of the delegations by recruiting better-qualified staff, primarily from producer organizations. The ministry intends to expand the activities of provincial delegations to include a much stronger emphasis on information flow. This will include reinforcing the information technology capacity of the delegations and organising a series of meetings with all actors in the sector in each province.

The information technology focus will probably be greeted with scepticism. At the provincial and municipal levels, most actors state that market information is not a panacea. A considerable amount of market information is already distributed both governmentally and to NGOs, who express doubts about its usefulness to their work. In isolated areas, where there is little access to markets, mere supply of information cannot solve the main constraint of access. This raises questions about the hopes expressed internationally that information technology will become a driving force in agricultural commercialization. As of yet, there are no programmes operational to test this hypothesis, but some will certainly come on-line in the near future.

Provincial MAGFOR delegations usually lack resources to mobilize their existing staff (jealousy towards the better-funded INTA structures is apparent). Programmes, such as control of vampire bats, that are designed to be ongoing are managed sporadically, depending on the availability of project funds. Crosscutting issues linking sanitation and human health are acknowledged to require better co-ordination with health institutions, but are currently addressed only on an *ad hoc* and occasional basis.

There is considerable interest within MAGFOR to strengthen governmental structures for technical education in agriculture, though the strategy is still unclear. In the meantime, the number of students in technical education in general and agriculture in particular is declining (UNDP 2000).

Non-governmental organizations have major roles in direct provision of extension services. The majority of extension workers in Nicaragua are probably employed by NGOs. Agent-to-farmer ratios are very high. Costs are invariably met through project aid. Sustainability and continuity are major problems, and in the mid-1990s, many NGOs were experiencing a crisis due to declining aid flows after the earlier post-conflict donor generosity. The influx of funds after Hurricane Mitch provided breathing space for many (Levard and Marín 2000), but the financial squeeze can be expected to return again in coming years. The NGOs have shown little interest in entering the market that is being created for private extension provision within the World Bank supported INTA programme.

Agricultural efforts of NGOs are focused on soil conservation, home gardens, and commercialization. They often provide extension combined with credit programmes, as capital is assumed to be a greater constraint than technological knowledge. Soil conservation and natural resource management programmes vary from short-term, food-for-work initiatives, wherein extension is a small add-on activity, to longer-term watershed management projects, often implemented in collaboration with the Ministry of Natural Resources and Environment. Home gardens are promoted partly as initiatives to support gender equity, and also as a means of diversifying income and diets. The NGOs take greater account of agriculture–health linkages than governmental agencies. Some of them even produce traditional medicines. Commercialization is an increasingly important theme for NGO extension efforts, and staff often display a strong awareness and concern for market factors. Exchange of experience among farmers is a major way that commercialization and technological assistance are generally supported.

Farmer-to-farmer approaches are well entrenched among many NGOs. The most prominent initiative is the farmer-to-farmer organization (*Programa Campesino a Campesino*, PCAC) within the (Sandinista-backed) National Union of Farmers and Ranchers (*Union Nacional de Agricultores y Ganaderos*, UNAG). This organization primarily promotes watershed management, sloping agricultural land technologies, home gardens and alternatives to 'swidden' agriculture (a type of slash-and-burn) through concrete extension projects as well as advocacy. It is well established and receives broad donor support. It also collaborates with international NGOs and research institutions, for example the International Centre for Tropical Agriculture (*Centro Internacional de Agricultura Tropical*, CIAT). Due to its Sandinista affiliation and outspoken criticism of conventional agriculture, it has limited collaboration with the government.

The projects run by PCAC and other NGOs with similar methods and goals have succeeded in establishing a certain level of national debate on alternatives to conventional agriculture. Some doubts exist about the longer-term financial viability of these types of extension programmes. The agricultural technologies themselves may perhaps be profitable. Critics point out, however, that rhetoric about farmers helping one another may hide a considerable level of donor-funded investment in extension staff and logistics. If these approaches are to become mainstream alternatives to conventional extension programmes it is essential that they are subjected to the same scrutiny as other types of efforts. That said, the costs of farmer-to-farmer approaches could be justified by the chance to reduce levels of environmental destruction. This comparison of costs and benefits will need to be made in a transparent manner, with an acknowledgement that environmental protection has an intrinsic value that may justify a significant level of subsidization.

Bilateral and multilateral projects usually employ their own extension personnel, often at wages far above what others provide. This has proved unavoidable in a context where bypass structures are virtually the norm, but it has severely distorted the labour market and incentive structures for technical assistance. The Swedish International Development Co-operation Agency (Sida) is planning to avoid this tendency by channelling its extension support through existing institutions.

Producer organizations, particularly in coffee, livestock, and nontraditional products, are getting increasingly involved in extension services, both as a part of their regular activities, and through projects. Services are not always limited to members. This is a positive development for their potential access by the poor, who are rarely active members of such organizations. On the other hand, such services are an indication that these agencies are either being pulled into the prevailing contract culture, or are acting as commercial service providers. In both situations, accountability to the organization's members is in danger of becoming a secondary priority.

Private extension service providers consist of technical assistance firms and individuals contracted directly by farmers or banks. The market for technical assistance firms was created largely by the establishment of INTA's private technical assistance facility, and will presumably grow if the vision of a gradual shift to contracting-out in the World Bank supported MAGFOR programme is expanded and becomes a national policy.

A number of individual private extension agents are active in providing services to wealthier farmers. Some banks demand that loan recipients, particularly for coffee, contract such individual extension providers as a way of reducing risk. A few individuals also provide some *ad hoc* training to groups of farmers, either on demand or in combination with input marketing.

Across the full range of extension providers there is scant capacity for broad strategic thinking or for monitoring and evaluation (Levard and Marín 2000). This is directly related to the supply and project driven nature of extension provision. Institutions are atomized and poorly articulated (Báez and Baumeister 1997). Continuity, efficiency, and equity are the victims of Nicaragua's projectized aid and extension market. Service provision is patchy. In some places (favoured by the so-called 'CNN effect' of media attention, drawing inappropriate concentration of resources) and with some technologies (home gardens), agencies are competing with each other to provide subsidies. In other areas (particularly inaccessible areas in the north and east), and for other farmer needs, there is a dearth of service providers. The MAGFOR recently put forth proposals for mobilizing 4000 extension agents which reflects an awareness of this problem; and despite concerns about the realism of the scheme, may nonetheless serve to stimulate a broader national discussion on the gaps in service provision.

## MAKING AGRICULTURAL EXTENSION AND RURAL DEVELOPMENT PRO-POOR: OPPORTUNITIES AND CONSTRAINTS

### THE POLICY FRAMEWORK

Extension priorities can be divided into two main categories in relation to livelihoods: helping poor people cope with their vulnerability, and helping them to 'escape' from poverty and thrive. The former emphasizes security, subsistence, and safety nets and the latter, commercialization, market participation, and increased income.

Internationally, the vast majority of governmental and commercial extension schemes have been justified on the basis that they contribute to thriving. The need to show a positive internal rate of return on investment has meant that thriving is in many cases taken for granted as the *raison d'être* for extension. Analyses of poverty, vulnerability, and nutrition all clearly point to thriving strategies as being most effective and 'sustainable' with respect to recurrent costs, dependency, and a limited role for public finance. Thriving is also increasingly dependent on information flows, but not on traditional technology transfer. Farmers need to understand and follow changing markets. They must also adapt to increasingly onerous sanitary controls in order to access international markets (Henson and Loader 2001) (see also Box 5.3 in relation to livestock products).

---

Box 5.3

The Emerging Market for Livestock Products in
Central America

Beef has traditionally been one of Nicaragua's main exports. Extensive production dominates, primarily in the agricultural frontier. Particularly during the period of 1960–90, vast areas of land were cleared, first for staple production, and then for cattle. Generalizations are difficult, but a cycle of development can be discerned whereby the agricultural frontier was initially colonized by small-farmers producing staples, followed by a deterioration of soil quality. Since poor producers lacked capital to convert their production to livestock, land holdings shifted to large-scale cattle ranchers (Maldidier and Marchetti 1996). In the older agricultural frontier, this has in many cases stabilized in a somewhat extensive but generally environmentally sustainable production system, combining dairy and meat. In the newer areas, there is an apparent shift from very extensive systems to near abandonment. Poor infrastructure and lack of capital have reduced beef production. There are vast areas of poor quality pasture with few or no animals. These areas are some of the poorest in the country. If livestock development could be revived, it could seemingly be a starting point for improving the livelihoods of the poor, even if it is indirect *via* the creation of employment, due to the current concentration of land ownership.

Most cattle ranchers combine milk and meat production, using milk to cover running costs and the sale of meat to generate profit. Since smaller producers require relatively regular income, they concentrate more on milk production. Larger milk producers employ a significant amount of wage labour, thus also having positive effects of poverty alleviation (Fernandez and Scoffield 2001). Infrastructure and access to dairy markets determine the relative balance between milk and meat. Labour intensive milk production is more attractive in the relatively accessible areas. Where new roads have been constructed, there is often a consequent increase in milk production, together with a general shift to more intensive production methods. Without infrastructure, there is little motivation for intensification and dairy production. Assumptions that industrial milk processing will overwhelm local and small-scale processing have thus far proven unfounded. Industrial production has stagnated at 20 per cent, and is concentrated in the most accessible areas. At the same time, small-scale dairy production and traditional cheese manufacture has expanded rapidly, now accumulating 60 per cent of national milk production (Cajina et al. 2000). Cheese is generally produced in areas with moderately poor infrastructure. If the infrastructure is very good, cheese producers must compete with industrial purchasers. If it is too poor, transport costs become too high. This implies that efforts to link extension to systems of collection and processing would be most effective if linked to these small 'semi-isolated' units, particularly as targeted to poor producers.

The major outlet for milk and cheese in the North has been the Salvadoran market, where milk prices are on an average 70 per cent higher than those in Nicaragua. The price paid for milk by Salvadoran traders is somewhat lower than by industrial purchasers, but the quality demands are also much lower making this market more attractive for the poor (Lorío 2001). There were fears that this market would shrink in 1999 after El Salvador imposed a ban on imports of (unpasteurized) products from uncertified plants, but since then the trade has continued unabated on an illegal basis, still handled primarily by Salvadoran traders.

This raises significant ethical questions regarding extension strategy, especially for poor producers who have the least potential to establish competitive and viable systems to pasteurize milk. Should an investment be made to strengthen a 'black market', even if it is undoubtedly an attractive market for the poor? If the market does not demand quality control, should extension priorities weigh the health concerns of importing and domestic consumers against the well-being of poor exporting producers?

The efforts of NGOs and food security programmes have more usually emphasized 'coping', as have many projects initiated after major crises. This alternative set of priorities is based on the belief that thriving will not reach everyone. Thriving is contingent on the availability of roads, markets, and institutions. It will not address the need to support livelihoods where social, economic, and physical infrastructures are not in place. Among neo-liberal Latin American economists, there is a growing readiness to assume that a significant proportion of rural peasant production is simply not viable.

This classification is becoming more common in referring to marginal areas in Latin America (Bebbington 1999). It is becoming acceptable not to invest limited finances in these areas, as people are assumed to be better-off migrating or finding different livelihoods, rather than remaining on their failing farms. Technocrats have often assumed that by merely ignoring the 'non-viable' communities, they will dissolve and join the mainstream.

However, this is not happening in Nicaragua. Instead, a destitute, alienated, and often-violent culture is becoming entrenched. The isolated agricultural frontier, where the forest has been cut to provide more agricultural land, has traditionally been the area that absorbed the poor and landless from the rest of the country. With the destruction of the forest nearly complete, this is certainly a less viable option than it was in the past (as can be seen from the poverty statistics), but there is no indication that these areas are being abandoned. A realization is emerging that coping strategies need to be supported, even if the mechanisms to support such strategies are not necessarily 'sustainable'. Market solutions alone will not lead to inclusive development. The internal rate of return on extension for isolated subsistence producers will probably be negative, and the prospects for significant cost recovery are nil, but these arguments are not sufficient to write these areas off. A mix of different strategies, most of which may require some form of subsidy (at least at the beginning) are needed.

In Nicaragua, the simple thriving–coping dichotomy does not fit neatly with real life livelihood strategies. It is nonetheless a useful heuristic device for relating extension to livelihoods. It should be noted that for poor people themselves, thriving and coping strategies are always entangled. Their use of assets to escape their current situations, and to survive in the meantime does not sort well into such categories. Furthermore, some of the labels for such strategies cause confusion about the coping–striving continuum. 'Diversification', for example, means very different things to poor farmers and to policy makers. Poor people diversify their livelihood strategies as a risk-reduction measure, by not putting all their eggs in one basket. In the policy discourse, diversification tends to mean that the country as whole should better distribute its eggs. Risks to the national economy could be mitigated by developing products for new markets. In order to enter non-traditional markets, individual farmers will inevitably need to specialize more (and diversify their household production less), thereby increasing their risk at the household level.

Suggestions are often made that niche products for export are a potential option for the poor. Small-scale producers, however, usually access markets via a learning process that begins with local markets, and then continues to national, regional, and international markets. With niche products, there is rarely a local market to use as a stepping-stone. Knowledge of international markets is limited among all types of extension staff, and among producers themselves. Risks are also very high, particularly for a small and disadvantaged (in terms of infrastructure and capital) country such as Nicaragua that has great difficulties competing with its neighbours. Given these risks, the poor are in many cases more likely to benefit from niche products, through employment generation effects, on medium and larger farms that can afford to take such risks.

One of the biggest niche products in Nicaragua is organic coffee. Poor producers are expected to draw benefits from organic production if transaction costs, specifically for certification and marketing, can be reduced to manageable levels. The price differentiation between organic and non-organic coffee is currently very wide, with non-organic coffee being mainly unprofitable at current prices. The NGOs support organic coffee production by subsidizing the initial period of learning and establishing routines of certification and marketing, thereby only burdening producers with the running costs of systems already in place. Some NGOs think it is essential that 'the producer must know the buyer', both for certification, and to understand the broader demands of quality control.

Many NGO efforts and food security programmes emphasize 'coping' strategies. This is due to both normative objectives and because projects were often initiated after major crises. A realization is emerging that coping strategies for those that lack these prerequisites need to be supported, even if the mechanisms to support these strategies are not necessarily 'sustainable'. Market solutions alone will not lead to inclusive development. A mix of subsidized and unsubsidized strategies is needed.

Despite relatively abundant and fertile land, Nicaragua has a major food deficit. Production of cereals has increased over the past decade, but at a cost of unsustainable conversion of forest and grassland to agriculture. Pessimism prevails about the capacity of Nicaraguan farmers to compete in the production of the primary staple—maize. The areas of the country with the highest per capita levels of food production are also those with the highest levels of poverty and malnutrition. Cereal production is the most common form of agricultural production by the poor. Food accounts for 60 per cent of expenditure of rural families, and malnutrition is highest among children in rural areas. These factors point to several difficult but fundamental questions. Should subsistence and cereal production be improved, or should alternatives be found? Should the emphasis on supporting poor people's livelihoods be on stimulating production (perhaps through higher prices) or entitlements for consumption (through lower prices)? The discourse on the future of subsistence farming and cereal production in Nicaragua is deeply divided. This is a part of the broader question of whether current livelihood strategies should be fortified, despite grave concerns about an inevitable decline in competitiveness, or whether farmers should be encouraged to abandon current priorities to invest in higher-risk alternatives.

Nicaragua is one of the most disaster-prone countries in the world. One result of this is that relief and rehabilitation programmes and social funds are (and deserve to be) a regular feature of the institutional landscape. Many NGOs involved in rural development started their programmes as part of post-war resettlement and other rehabilitation projects. It is within such schemes that some of the most positive examples have emerged of reconciliation in a country that is otherwise torn by polarization. Little systematic attention, however, has been paid to finding and developing synergies between these projects and long-term development programming. Moreover, with the notable exception of some watershed management and soil conservation efforts, there have been relatively few attempts to address disaster risks in development planning. There is a significant role for extension in addressing the issues of: (i) increasing the impact of rehabilitation efforts on long-term development and (ii) increasing the impact of long-term development efforts to reduce risk and vulnerability. The frequently poor performance of these programmes is not solely due to the short-sightedness of the planners of emergency and rehabilitation programmes. A major problem has been the lack of readiness of development planners to look for ways to integrate and utilize these efforts in their programming. Extension staff are at the frontline of these processes, and can be expected to play a key role in addressing this gap.

Watershed management and related interventions to improve land husbandry on sloping land receive considerable attention. Before Mitch, this was justified on environmental grounds, to mitigate the environmental destruction underway at the agricultural frontier and to intensify resource use in order to reduce pressures for further expansion into the rainforest. After Mitch, two additional justifications came to the forefront. These projects are now being promoted on an expanding scale as ways to reduce the risk of disasters (especially landslides) and as windows for safety nets (food or cash for work). However, many extension staff remain highly sceptical of such schemes, seeing them as exceedingly staff-intensive and expensive.

## LIVELIHOODS AND EXTENSION IN NICARAGUA

Poor people in rural areas are producers, consumers, labourers, and citizens. Technological change affects them differently according to these different roles. Promotion of technological change in agriculture will impact on the lives of the Nicaraguan poor through greater entitlements in the form of three overlapping categories: (i) production and labour markets, (ii) reduced vulnerability, and (iii) greater empowerment.

### Production and Labour Markets

Entitlements can be enhanced through increased production/productivity and access to employment. Basic elements of increasing production and enhancing labour markets through technological change include the following:

1. increased cereal production for consumption and commercialization;

2. diversified diets, primarily through home gardens and small stock;

3. taking advantage of new commercialization opportunities, particularly in conjunction with access to expanding infrastructure (example, dairy);

4. improved marketing and 'good exits' from agriculture through an invigorated rural service sector, including processing and small enterprise development;

5. labour-intensive production technologies on larger farms to create employment;

6. labour-saving technologies for small-scale producers to increase competitiveness and opportunities for diversification;

7. programming that builds on the relationships between labour markets and harvesting/processing technologies;

8. intensification to make greater and more efficient use of family labour; and

9. skills for migrants and semi-skilled agricultural labourers.

## Reduced Vulnerability

Vulnerability reduction involves increased resilience to livelihood shocks, environmental protection, access to safety nets, and better health and nutrition, that is, addressing the myriad risks that confront poor and better-off households.

Examples of vulnerability reduction priorities include:

1. enhanced environmental health by reducing pollution from processing facilities, and more appropriate use of agro-chemicals;

2. better nutrition through cheaper, more varied and nutritious (and even medicinal) diets;

3. access to safer foods (especially dairy);

4. reduction of production risks through lower-risk technologies;

5. diversification of on-farm and off-farm asset investment;

6. reduction of risks of landslides, erosion, etc.;

7. enhanced community/household food security through greater access to entitlements in the event of livelihood shocks, including making the best of post-disaster safety nets, such as cash/food-for-work programmes;

8. improved quality and impact of rehabilitation projects through better links to development strategies;

9. insurance; and

10. mitigation of rural violence through livelihood opportunities for the youth and marginalized groups.

## GREATER EMPOWERMENT

'Poverty is related to the lack of political power of the poor' (Government of Nicaragua 2000), and inevitably the poor will need a stronger stance in dealing with institutions of government and the market if they are to transform production increases into better livelihoods. Power is related to knowledge of the market for their products, the ability to update that knowledge, and institutions that create a critical mass for negotiation and a choice of production options. Extension can deal with some of these factors directly. In others, its role will need to be developed within a broader policy and institutional environment that enhances the power of poor people to exert their demands. Education is the single most important factor in improving the welfare of rural households (World Bank 2000), and it is, therefore, imperative that extension strategies are formed in relation to an overall focus on knowledge as the linchpin of rural development. There are six areas where such empowerment can be promoted:

1. skills that increase the producer's power to negotiate (knowledge of marketing, quality control, certification bureaucracies, etc.);

2. infrastructure that increases the producer's power to negotiate (storage, processing, and other post-harvest technologies);

3. organizations that increase the producer's power to negotiate and demand services;

4. the existence of more than one person with whom to negotiate (that is, more traders/competition and a more dynamic service economy);

5. control ôf the production process through producer capacity to manage linkages of credit, processing, marketing, quality control, and input supply; and

6. diversification to avoid dependence on one crop/buyer/processing structure.

## CONCLUSIONS: REFOCUSING PRIORITIES

What should be the link between agricultural (and rural development) policy and existing survival strategies? If poverty is to be addressed by extension in Nicaragua, a two-phase approach is needed, drawing on different geographic priorities and potentials.

HIGH POTENTIAL AND ACCESSIBLE AREAS

Such areas include:

1. commercialization of fruit, vegetable, livestock, and dairy production;

2. expanded irrigation;

3. labour-saving technologies for household production;

4. labour-intensive technologies for large-scale production;

5. environmental health interventions;

6. improvement of production quality, timeliness, and sanitation; and

7. targeting extension inputs to areas made accessible by new infrastructure.

In these areas, thriving should be the major focus for both direct and indirect (wage labour) opportunities. There should be preparedness, however, to support coping strategies when required, as thriving carries with it some increased vulnerabilities. The private sector dominates the agenda for technological change in accessible areas and supermarkets play an increasing role in product marketing (Box 5.4). The public sector has a relatively limited role, and should emphasize clearly defined public goods, especially those related to health, sanitation, and nutrition. Labour markets should be a major factor in programming, albeit with an acceptance of the fact that government policy can influence but presumably not lead developmental trajectories. There is also a role for the public sector to provide

---

BOX 5.4

Supermarkets and the Poor

Rural development planners often implicitly assume that supermarkets are a relatively small market, supplied by a few wealthy producers for a few wealthy consumers. According to this logic, this 'niche' is of little relevance for a discussion of pro-poor extension. Realities are changing. In Nicaragua, the supermarket share of the food market is 15–20 per cent and the number of supermarkets has grown from 5 to 60 in less than ten years. In most Latin American countries it is far larger

(Reardon and Berdegue 2002). There are strong indications that the poor will either need to start accessing supermarkets (in Nicaragua or abroad) or see their share of markets decline rapidly.

Selling to supermarkets requires close supervision and market information flow to ensure quality, timeliness, and transport. The NGOs and producer organizations are establishing collection and processing centres that provide packages of extension, inputs, processing, and marketing. These schemes support mainly vegetable production. The agencies managing such efforts also take care of arranging contracts with buyers, such as national and regional supermarket chains and other large-scale consumers including hospitals and hotels. In one case, a producer of organic vegetables for export to the United States (one of the first) is offering similar services to nearby farmers, renting the services of a producer organization's processing facilities. In these examples, high levels of extension inputs have proved essential for maintaining quality (both in the field and in post-harvest processing) and also for ensuring that products are available by market demand. Timing is extremely important in vegetable production, both to maintain buyer confidence and to avoid flooding the market.

Such schemes have been directed primarily towards small-scale farmers using irrigated land. While relatively poor, access to irrigated land is an indication that these producers are not among the very poor. Programmes such as these do not often have explicit poverty alleviation objectives. Increased income among those moderately well-off farmers who participate is the main objective. Large-scale producers are increasingly seen as essential to open markets and ensure that sufficient bulk produce is available to attract major buyers.

The activities of many of these commercialization centres were originally financed with aid resources, but unlike many other NGO efforts, organizations explicitly strive for financial viability (even with the costs of extension included). Public finance for infrastructure and organizational support appears to be essential in the early stages of establishing commercialization programmes directed at small-scale producers, as private capital will otherwise go towards large-scale producers. Collaboration between producer organizations and NGOs seems to be most appropriate, with NGOs initially providing aid-financed extension, before turning it over either to the producer organization itself, or to the firms handling commercialization. Extension costs are eventually included in producer prices.

For both contract farming and other commercialization schemes, the only opportunity for financially viable small-scale farmer participation is through strengthening producer organizations, either pre-existing or new.

technical back-up to re-establish production after a disaster, where the private sector is overwhelmed, and where capital is in short supply.

LOW POTENTIAL AND ISOLATED AREAS

These areas include:

1. products with high value relative to transport cost;

2. diversification of diets;

3. focus on areas that will soon become accessible with new infrastructure;

4. subsistence production;

5. natural resource and watershed management;

6. more effective use of safety nets; and

7. skills for migration.

Coping strategies will dominate the agenda for low potential and isolated areas, although some openings do exist for limited thriving strategies. While there is a great need for investment in extension in these areas, it is doubtful that the public sector will be able to cover the level of recurrent costs for services that will reach the diverse and scattered populations of the distant agricultural frontier. It has been noted that there is a global trend for states to abandon areas such as these to non-state actors from the private sector, civil society, and even uncivil society (Duffield 2000). To suggest that public sector extension should buck this overall trend is rather over-optimistic. There is, however, some potential for public service institutions to be contracted in, that is, to be used by aid projects to engage in tasks for which they otherwise lack resources, while providing a skilled, knowledgeable, and locally based organization to contract agencies. Rehabilitation programming is an obviously important window for such contracting in. Experience has shown that this regrettably leads to 'adhocracy' and the need to address another layer of problems— that of finding synergy between rehabilitation and development. Links between temporary safety nets and 'normal' development should not be assumed to be inherently dysfunctional.

THE 'END OF THE ROAD'

Within this dichotomy between dynamic areas and areas that are perhaps out of reach of weakened state institutions, there is also a third discernible set of targets—'the end of the road'—where infra-structure is improving, but where market forces are not yet fully established the United Nations Development Programme (UNDP) states that 'territorial integration is a fundamental step in social and economic integration' (UNDP 2000). Roads have a major impact on technological change, so it is imperative that extension takes this factor explicitly into account. Roads create threats and opportunities. They may: (i) accelerate destruction of the forest; (ii) raise land values and, therefore, encourage more sustainable land husbandry; (iii) raise land prices, forcing the poor to sell their land; (iv) encourage investment by large-scale producers that creates employment oppor-tunities; (v) encourage investment in capital-investment technologies, displacing the poor; (vi) open access for low-priced imports to compete with existing production; and (vii) create access to markets for poor producers.

Together, these and other factors create a complex and dynamic mix of pressures on poor people and on agents for technological change. A central challenge for extension is to monitor and adapt to this mix of opportunities and threats. Nicaragua's road network is expanding, and producers are adapting to the new opportunities and threats that these roads represent. In deciding how to most effectively employ a few hundred extension agents (the scale of INTA's opera-tions), targeting areas where new infrastructure is just opening opportunities for commercialization and income enhancement would seem an obvious priority. There may be particular synergy where soil conservation and watershed management efforts on higher-potential sloping land suddenly offer possibilities for better links to markets. There is also an increased need for risk-mitigation efforts, as roads lead to increased deforestation and may also be designed with insufficient regard to gully formation and landslide risk.

## PRO-POOR EXTENSION AMIDST POLITICS AND POLICIES

Extension practice is derived from a mix of incentives, regulations, relationships, and visions. Ideally, a democratic political process should define parameters that are then codified in policies, to inevitably guide practice, often with the support of projects. This is often not the case in Nicaragua, where the interplay between these projects and politics tends to outweigh the influence of a consistent political vision in guiding policy formation for extension practice. Local politicians derive their prestige, legitimacy, and often their identity from bringing projects (especially visible infrastructure) to their constituents (see Larson 2001; Tendler 1997). With such a state of affairs, it is easy to become cynical about the scope of policy led agricultural development.

Cynicism regarding the role of agricultural extension derives from the fact that extension, as the bearer of technology, is naturally assumed to be a set of institutions that should fit hand-in-glove with a technocratic vision of development. When placed amidst the messiness of Nicaraguan policy on rural development and poverty alleviation, disillusionment easily sets in. As Tendler (1997) has pointed out, however, this type of situation is not as grim as it seems. It is possible to find ways of linking to ongoing processes of political and institutional change that may create openings for state institutions, civil society, and the private sector to reach poor farmers. Common interests can be found, even if the path to finding such interests fails to resemble a linear model of policy implementation. In order to find the levers that relate extension reform to poverty reduction, it is important to accept that they will not always emerge from an overall policy vision, but may often be found in a the minutiae of local process.

A new narrative of policy formation that retains poverty alleviation objectives is needed, while putting aside hopes that extension will pick the 'right' technologies. It should be asked: have extension actors expanded (or shown the potential to expand) the range of choice and options of the poor in using their resources; are the poor expanding their market involvement and thereby their exposure to risk; or are they 'hunkering down' with subsistence and on-farm diversification; who has/could help them pursue either strategy more effectively? These questions relate to how the importance of vulnerability is perceived. Should extension (and the policies in which it is embedded) seek to help farmers find more baskets in which to place their eggs? Should extension support be directed at helping farmers to manage a broader and more flexible portfolio of investments, on and off the farm? Or, should extension help farmers feel secure enough to transcend the 'egg basket' vulnerability paradigm and plunge into intensive, full-time, competitive commercial production, using other mechanisms (for example, insurance or safety nets) to cushion the increased risk? (see Figure 5.1).

Triage is a useful concept for facing the questions surrounding extension and policy formation. Triage is defined as 'the principle or

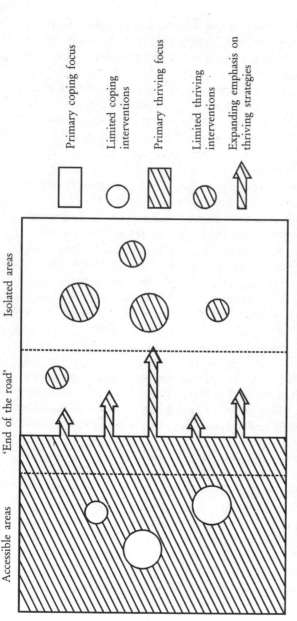

FIGURE 5.1: Schematic view of extension strategies in relation to degree of market integration

*Note:* Extension strategies should differ considerably according to the level of market integration. The primary focus in reaching poor farmers in isolated areas will inevitably be on support to coping strategies, though there will even be some limited possibilities to encourage commercialization. In accessible areas, the emphasis will be on thriving strategies, although there should also be readiness to help farmers in these areas to cope with livelihood shocks. As areas become accessible with new infrastructural development ('the end of the road'), extension has an important role in helping the poor to take advantage of new, emerging market opportunities.

practice of allocating limited resources, as of food or foreign aid, on a basis of expediency rather than according to moral principles or the needs of the recipients' (Collins 1991). This usage of the term stems from battlefield medicine, where casualties are sorted according to those who will survive without treatment, those who will probably not survive at all, and those in-between for whom treatment will yield greatest impact. Even though triage is a word rarely used in studies of extension, it has nonetheless been a guide for many extension investments. It is a useful way of shedding light on the practical and moral choices to be made in extension prioritization, and for placing this prioritization within the broader context of rural development policy.

The recommendations presented here suggest that extension directed at producers themselves will yield diminishing returns (relative to costs) with isolated and very small producers. If we ask whether or not small-scale production for the poorest is a worthwhile investment, we open the door to better differentiating between actions that have direct, indirect, or improbable/undefined impacts on the poor. It is also a useful concept for specifying how far down the poverty line one can hope to reach with a given type of intervention. The two zones mentioned above, plus the end-of-the-road target area, provide a graphic structure for sorting through these choices.

It may be that interventions with larger-scale producers have a better chance of impacting the poorest than support to relatively small-scale farmers. The latter depend primarily on household labour, while larger-scale producers employ landless labourers and provide the extra income that is required for a household to stay on in the rural areas. Small-scale coffee producers, for example, are not among the poorest of the poor. They usually farm using almost exclusively family labour. It may actually be the case that in order to produce a labour market for the landless, it would be better to target larger-scale producers, as it is they who employ a significant number of wage labourers. A symbiotic relationship can be found between impoverished rural households in Nicaragua and the use of labour-intensive farming technologies on large farms in Costa Rica.

In using analytical frameworks such as this, triage highlights a number of difficult policy trade-offs that are rarely addressed in extension planning. As costs rise relative to production benefits with small-scale or isolated producers, the question becomes one of the relative appropriateness of different subsidies (for example, between subsidizing input supply, marketing, organizational support, or finance). Each of these types of subsidies has emerged over the years as a panacea for inclusive rural development. Assumptions that subsidies can merely be withdrawn after a few years when everything has become 'sustainable', have either proved false, or convinced planners that these programmes must be redirected at a somewhat wealthier target group. Calls are emerging to reassess standard 'rules' about donor funding and recurrent costs (Arana *et al.* 1998), but this can only be done if the broader issue of safety nets is taken out of the sustainability closet.

Trends in rural development in the face of globalization have shown that this issue is more acute than ever. 'Durable disorder' (Duffield 2000) is now taking hold in marginal areas in the form of chronic violence and social alienation. Transnational economic networks are taking advantage of the withdrawal of the state from isolated rural areas by establishing smuggling, production of narcotics, and other forms of illicit enterprise. This phenomenon suggests that there are heavy economic costs (in addition to moral issues) stemming from conflict, criminality, and social disintegration when services are withdrawn.

How might a pragmatic approach be achieved, accepting that Nicaragua is a land of projects, but rejecting the tendency to put poverty on the back-burner? If triage is to be used as an analytical concept for understanding these choices, and not as a recipe for exclusion, then greater moral and pragmatic articulation between policy formation and programming is needed. This, in turn, will mean bringing agriculture into the sphere of national debates on poverty and the government–civil society–donor triangle that emerged after Hurricane Mitch. It is only if all three of these actors take their national policy commitments more seriously, as guides for operational planning, that the current fragmentation can be overcome. Much of the potential for using extension as a tool for poverty alleviation currently falls between the cracks of rural development programming. This is particularly true of vulnerability, since alternative models for growth, from both left and right, have shown a repeated tendency to ignore, and thereby aggravate, risk.

*References*

Arana, Mario, Juan Sebastián Chamorro, Silvio de Franco, Raúl Rivera, and
    Manuel Rodríguez (1998), 'Revisión del Gasto Público en Servicios

Sociales Básicos: Financiamento, Equidad, Eficiencia e Impacto, el Caso de Nicaragua', (Public Expenditure Reform in Basic Social Services: Finance, Equity, Efficiency and Impact, the Case of Nicaragua) (mimeo).

Ardón, Patricia (1999), 'Post-war Reconstruction in Central America', Oxfam Working Papers, Oxford, UK: Oxfam.

Arróliga, Lourdes (2001), 'Fracasa Gran Proyecto de Riego', (Grand Fiasco in Irrigation Project), *Confidencial*, Vol. 5, p. 231, 4–10 March 2001, *http://www.confidencial.com.ni/deportada.html* 2001-03-09

Báez, Lacayo, and Linda and Eduardo Baumeister (1997), 'Estudio sobre Pobreza Rural y Alternativa de Desarollo para Pequeños Productores', Consultants report for Swedish International Development Agency (Sida), Stockholm, Sweden: Sida (mimeo).

Barandun, Armin (2001), 'Developing a Mix of Public and Private Rural Extension Services: The Experience of Nicaragua', LBL Landwirtschaftliche Beratungszentrale, Swiss Centre for Agricultural Extension (LBL) (mimeo).

Bebbington, Anthony (1999), 'Capitals and capabilities: A framework for analysing peasant viability, rural livelihoods and poverty', *World Development*, Vol. 10, No. 1, pp. 17–34.

Berdegué, Julio, Thomas Reardon, Germán Escobar, and Rubén Echeverría (2000), 'Policies to Promote Non-farm Rural Employment in Latin America', *ODI Natural Resource Perspectives* 55, London, UK: Overseas Development Institute (ODI).

British Red Cross (1999), 'Hurricane Mitch Agricultural Support Programme: Rebuilding Rural Livelihoods, a Comparative Analysis of the Effects of Cash Assistance on the Recovery of Rural Livelihoods', London, UK: British Red Cross.

Brown, Adrienne, Mick Foster, Andy Norton, and Felix Naschold (2001), 'The Status of Sector Wide Approaches', ODI Working Paper 142, London, UK: Overseas Development Institute (ODI).

Cajina, A. J., R. Blandón, J. Díaz, and G. Mora (2000), 'Documento base de Discusión Sobre la Situación del Sector Lácteo de Nicaragua' (Discussion Paper on the Dairy Sector in Nicaragua), Managua, Nicaragua: Federación Centroamericano y del Caribe del Sector Lácteo (Central American and Caribbean Dairy Federation) (FECCALAC).

Cheru, Fantu (2001), 'Economic Social and Cultural Rights: The HIPC Initiative: A Human Rights Assessment of Poverty Reduction', Strategy Papers, United Nations Economic and Social Council E/CN.4/2001/56, New York, USA: UNESCO.

Christoplos, Ian, J. Mitchell, and A. Liljelund (2001), 'Reframing Risk: The Changing Context of Disaster Mitigation and Preparedness', *Disasters*, Vol. 25, No. 3, pp. 185–98.

Collins (1991), *Collins English Dictionary*, Glasgow, UK: Harper Collins Publishers.

Corral, Leonardo and Thomas Reardon (2001), 'Rural Non-farm Incomes in Nicaragua', *World Development*, Vol. 29, No. 3, pp. 427–42.

Delgado, C., M. Rosegrant, H. Steinfeld, S. Ehui, and C. Courbois (1999), 'Livestock to 2020: The Next Food Revolution', Food, Agriculture and the Environment Discussion Paper 28, Rome, Italy: International Fund for Agricultural Development (IFAD).

Devereux, Stephen (2000), 'Making Less Last Longer: Informal Safety Nets in Malawi', IDS Discussion Paper, 373, Institute for Development Studies (IDS), Brighton, Sussex, UK: University of Sussex.

Dinar, Ariel and Gabriel Keynan (1998), 'The Cost and Performance of Paid Agricultural Extension Services: The Case of Agricultural Technology Transfer in Nicaragua', World Bank Policy Research Working Paper 1931, Washington, DC, USA: World Bank.

Duffield, Mark (2000), 'The Emerging Development–Security Complex', in Paul Collins (ed.), *Applying Public Administration in Development: Guideposts to the Future*, Chichester, UK: John Wiley and Sons.

Fauné, Angélica and David Kaimowitz (1999), 'Posoltega: La Necesidad de Articular Esfuerzos y Recursos en Torno a la Reforestación' (Posoltega: the Necessity to Articulate Power and Resources for Reforestation), Managua, Nicaragua: Nitlapán—Centre for International Forestry Research (CIFOR)—Proyecto de Municipales Rurales (Rural Municipality Project) (PROTIERRA).

Fernandez, Edgar and John Scoffield (2001), 'Informe de Consultoría sobre el Componente de Ganaderia' (Consultancy report on the Livestock Component), Matagalpa, Nicaragua: Orgut (mimeo).

Fitzgerald, Valpy and Arturo Grigsby (2001), 'Nicaragua: The Political Economy of Social Reform and Armed Conflict', in Francis Stewart and Valpy FitzGerald (eds), *War and Underdevelopment: Volume II Country Experiences*, Oxford, UK: Oxford University Press.

Frühling, Pierre (2000), *After Mitch*, Stockholm, Sweden: Swedish International Development Agency (Sida).

González, Mary Lisbeth (2001), 'Estrategias de Sobrevivencia de los Pobres en Nicaragua' (Survival Strategies of the Poor in Nicaragua), *Confidencial*, Vol. 5, p. 230, 25 February–3 March 2001, *http://www.confidencial.com.ni/2000-230/analisis.html*, 2001-03-09

Government of Nicaragua (2000), 'Estrategia Reforzada de Reducción de la Pobreza' (Strengthened Poverty Reduction Strategy), 18 May 2000, Managua, Nicargua: Government of Nicaragua.

Henson, Spencer and Rupert Loader (2001), 'Barriers to Agricultural Exports from Developing Countries: The Role of Sanitary and Phytosanitary Requirements', *World Development*, Vol. 29, No. 1, pp. 85–102.

Humphries, Sally (1998), 'Milk Cows, Migrants, and Land Markets: Unraveling the Complexities of Forest-to-Pasture Conversion in Northern

Honduras', *Economic Development and Cultural Change*, Vol. 47, No. 1, pp. 95–124.

Jiménez, Alí (1999), 'Nicaragua: Pobreza e Inseguridad Alimentaria' (Nicaragua: Poverty and Food Insecurity), consultant's report prepared for Ministry of Agriculture Livestock and Forestry (MAGFOR) Food and Agriculture Organization of the United Nations (FAO), Managua, Nicaragua (mimeo).

Larson, Anne (2001), 'Natural Resources and Decentralizations in Nicaragua: Are local governments up to the job?', *World Development* (submitted).

Levard, Laurent and Yuri Marín L. (2000), 'La Problemática Técnica y la Intervención de los Organismos que Trabajan en el Fomento Tecnológico en el Trópico Seco de Nicaragua' (Technical Issues and the Intervention of Organizations that Work in Technological Training in the Tropical Dry Zone of Nicaragua), *Encuentro*, Vol. 32, No. 53, pp. 11–18.

Long, Norman and Magdalena Villarreal (1998), 'Small Product, Big Issues: Value Contestations and Cultural Identities in Cross-border Commodity Networks', *Development and Change*, Vol. 29, pp. 725–49.

Lorío Castillo, Margarita (2001), 'Estrategias para Aumentar el Nivel de Ingresos y Mejorar su Distribución en Hogares de Pequeños Productores de Café y Leche en Municipios de Matagalpa y Jinotega' (Strategies for Increasing the Level of Income and Increasing its Distribution in Households of Small Producers of Coffee and Milk in the Municipalities of Matagalpa and Jinotega), Consultants report prepared for Matagalpa, Nicaragua. Orgut (mimeo).

MAG (1998), 'Una Ruta para Modernizar la Nicaragua Rural', Document prepared for the meeting of the consultative group on Nicaragua, Geneva, April 1998, Consultative Group on Nicaragua (mimeo).

MAGFOR (2001), 'Estrategies para el Desarrollo de la Agricultura Nacional Horizonte 2010' (Development Strategies for National Agriculture Towards 2010), Managua, Nicaragua: Ministry of Agriculture, Livestock and Forestry (MAGFOR).

Maldidier, Cristóbal and Peter Marchetti (1996), 'El Campesino-Finquero y el Potencial Económico del Campesino Nicaragüense' (The Peasant Farmer and the Economic Potential of the Nicaraguan Peasantry), Managua, Nicaragua: Nitlapán.

Marín, L., Yuri (2000), 'Limitaciones y Perspectivas del Cultivo de Plátanos (*Musa sp.*) en el Departmento de Rivas' (Limitations and Perspectives on the Cultivation of Plantains), *Encuentro*, Vol. 32, No. 53, pp. 30–9.

Martínez, S., J. Arnoldo (1999), 'Reflexiones sobre la Vulnerabilidad Nacional (Reflections on National Vulnerability), *Confidencial*, 128, 31 January–6 February, *http://www.confidencial.com.ni/l.../body-columnista-invitado.htm 1999-02-19*

Matin Imran, David Hulme, and Stuart Rutherford (1999), 'Financial Services for the Poor and Poorest: Deepening Understanding to Improve Service

Provision', Finance and Development Research Programme Working Paper 9, Manchester, UK: Institute of Development Policy and Management (IDPM), University of Manchester.

Nitlapán (2001), 'Programa de Investigación Octubre 1999–2002: Objectivos, Avances de los Años 1999 y 2000, Perspectivas para 2001 y 2002' (Objectives and Progress in the Years 1999 and 2000, Perspectives for 2001 and 2002), Mangua, Nicaragua: Equipo de investigación sobre Tecnología Agropecuaria Nitlapán (mimeo).

Nogueira, Roberto Martinéz (1990), 'The Effect of Changes in State Policy and Organization on Agricultural Research and Extension Linkages: A Latin American Perspective', in David Kaimowitz (ed.), *Making the Link: Agricultural research and technology transfer in developing countries*, Boulder, Colorado, USA: Westview Press.

Peppiatt, David, John Mitchell, and Penny Holzmann (2000), *Buying Power: The Use of Cash Transfers in Emergencies*, London, UK: British Red Cross.

Radix (2001), *http://www.anglia.ac.uk/geography/radix*

Reardon, T., and J. A. Berdegue (2002), 'The Rapid Rise of Supermarkets in Latin America: Challenges and Opportunities for Development, *Development Policy Review*, Vol. 20, No. 4.

Reyes, E. and R. Rodriguez (1998), 'Reforms in the Financing and Delivery of Agricultural Extension Services: Colombia', in Abigail Mulhall, Kate Warren, and Chris Garforth (eds), *Reforms in the Financing and Delivery of Agricultural Extension Services: Cases from Africa, Asia and Latin America*, AERDD Working Paper 98/3, Reading, UK: Agricultural Extension and Rural Development Department (AERDD), University of Reading.

Rocha, José Luis and Ian Christoplos (2001), 'Disaster Mitigation and Preparedness in the Nicaraguan Post-Mitch Agenda', *Disasters*, Vol. 25, No. 3 (in press).

Schafer, Jessica (2001), 'Supporting Livelihoods in Situations of Chronic Political Instability', London, UK: Overseas Development Institute (ODI) (mimeo).

Secretaría de Acción Social de la Presidencia de la República (Secretariat for Social Action of the President of the Republic) (2000), 'Política Nacional de Seguridad Alimentaria y Nutricional' (National Policy on Food and Nutritional Security), Managua, Nicaragua: Secretaría de Acción Social de la Presidencia de la República.

Secretaría Ejecutiva del Sistema Nacional de Prevención, Mitigación y Atención de Desastres (Executive Secretariat for the National System for Prevention, Mitigation and Attention to Disasters) (2000), *Ley 337: Ley creadora del Sistema Nacional de Prevención, Mitigación y Atención de Desastres*, (Law 337: Law Creating a National System for the Prevention, Mitigation and Attention to Disasters), Manaugua, Nicaragua: United Nations Development Programme (UNDP).

Tendler, Judith (1997), *Good Government in the Tropics*, Baltimore, Maryland, USA: Johns Hopkins University Press.

UNDP (2000), *El Desarrollo Humano en Nicaragua: Equidad para Superar la Vulnerabilidad*, Managua, Nicaragua: United Nations Development Programme (UNDP).

WFP (2001), *Country Strategy Outline—Nicaragua*, Rome, Italy: World Food Programme (WFP).

World Bank (2002), *2002 World Development Indicators* CD Rom, Washington, DC: World Bank.

World Bank (2000), *World Development Report 2000/2001: Attacking Poverty*, Oxford, UK: Oxford University Press.

# 6
# Extension, Poverty, and Vulnerability in Bolivia
*Alan J. Bojanic*

## THE COUNTRY CONTEXT

### BASIC INDICATORS

Bolivia's population is estimated to be around 8 million, of whom nearly two-thirds live in the highlands (*Altiplano* and inter-Andean valleys) and one-third in the lowlands. By 1997, nearly 59 per cent of the population was considered urban, compared to 40 per cent in 1976 (Pacheco 1998). Patterns of both rural–rural and rural–urban migration continue to provide an escape route from poverty in the rural highlands. At 2.2 per cent per annum, the population growth rate can be considered high, particularly in relation to other lower-middle income countries (1.2 per cent per annum). Other basic economic and social indicators are summarized in Table 6.1.

In eco-geographical terms, Bolivia is divided into three large sub-regions. These make up a total of 1,098,581 sq. km, an area almost twice the size of France (Montes de Oca 1997): (i) high plateau or Altiplano (surface area 246,254 sq. km); (ii) inter-Andean valleys (168,320 sq. km); and (iii) tropical and sub-tropical lowlands (684,007 sq. km).

### RURAL POVERTY

Compared to other South American countries, Bolivia still has the largest rural population and the highest percentage of people living below the poverty line (66 per cent in 1997, Morales 2000). In rural areas, poverty levels can be as high as 90 per cent. This grim picture also holds true for other social indicators.

TABLE 6.1
Bolivia, Basic Indicators

| Series | Value | Year |
|---|---|---|
| Cereal yield (kg per hectare) | 1676 | 2000 |
| Land use, arable land (hectares per person) | 0.24 | 1999 |
| Land use, irrigated land (% of cropland) | 6 | 1999 |
| Agriculture, value added (% of GDP) | 22 | 2000 |
| GNI per capita, Atlas method (current US$) | 990 | 2000 |
| Population, total | 8,328,700 | 2000 |
| Rural population (% of total population) | 35 | 2000 |
| Malnutrition prevalence, height for age (% of children under 5) | 27 | 1998 |
| Malnutrition prevalence, weight for age (% of children under 5) | 8 | 1998 |
| Low birthweight babies (% of births) | 9 | 1995 |
| Poverty headcount, national (% of population) | Not available | – |
| Poverty headcount, rural (% of population) | 79 | 1995 |
| GINI index | 44.7 | 1999 |
| Mortality rate, infant (per 1000 live births) | 57 | 2000 |
| School enrolment, primary (% net) | 97 | 1998 |
| Surface area (sq. km) | 1,098,580 | 2000 |
| Roads, total network (km) | 53,790 | 2000 |

Source: World Bank (2002).

The wide ecological diversity of the country is mirrored by its diversity in culture and farming systems. In Bolivia there are more than 40 indigenous groups who maintain strong cultural identities and still use their own language, customs, and traditions. According to the Bolivian Institute of Ethnology, more than 70 per cent of the Bolivian people are considered to be indigenous in origin, the highest percentage among Latin American countries. The spatial distribution of poverty is given in Map 6.1.

POLICIES TOWARDS AGRICULTURAL AND
RURAL DEVELOPMENT

At 15 per cent, the agricultural sector still maintains a high contribution to gross domestic product (GDP) and accounts for nearly 35

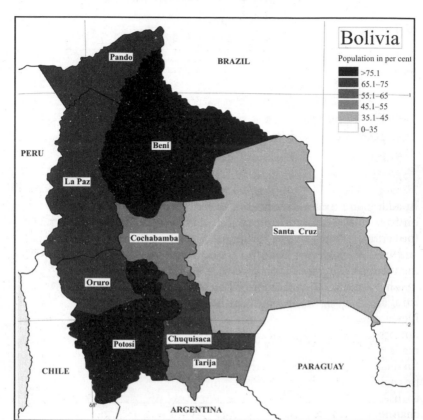

1994 MAGELLAN GeographixSM Santa Barbara, CA (800) 929-4627

MAP 6.1: Spatial distribution of persons living in poverty in Bolivia in 2001 by department

Source: Based on Table 1, 'Bolivia: Poverty Map, 2001', Bolivian National Institute of Statistics, at:

http://www.ine.gov.bo/PUBLICACIONES/Censo_2001/Pobreza/PBolvivia.pdf

per cent of all exports. Most agricultural exports are primary products, particularly soybean grains, cotton, sugar, and coffee. Imports of foodstuffs are high, particularly wheat and dairy products. Agriculture has expanded in the lowlands, encouraged by the high returns for soybeans, but little expansion has occurred in the valleys or Altiplano. Agricultural yields remain among the lowest in Latin America, particularly for staple crops.

The main problems or constraints facing the agricultural sector are:

1. fragmentation of land holdings and land-tenure schemes (*minifundio*);

2. harsh climatic conditions;

3. low productivity;

4. poor access infrastructure and long distances to markets;

5. soil erosion and pastureland degradation; and

6. lack of agricultural credit schemes.

In addressing these challenges, Bolivia bases its development strategy on four pillars: (i) opportunity (production-related issues); (ii) equity (poverty alleviation and provision of social services such as education and health); (iii) dignity (fight against cocaine production); and (iv) institutions (decentralization and popular participation).

The promotion of rural development and reduction of poverty in rural areas fall under the pillar of equity. Within the Comprehensive Development Framework, co-ordinated by the World Bank and donors interested in supporting rural development, efforts have been made to develop a common strategy and to co-ordinate the use of financial resources. An additional way the government envisages financing the implementation of anti-poverty programmes has been to use the highly indebted poor countries (HIPC) II[1] initiative to channel funds from debt relief to reduce poverty. The total external debt of Bolivia is approximately US$ 4.4 billion and it is estimated that up to 30 per cent of this (US$ 1.3 billion) could be released for poverty reduction projects (US$ 100 million per annum over 13 years) through the HIPC structure. A process of consultation has started in which most actors in the public sector have agreed to an agenda for spending such resources.

The main vehicle for reducing poverty in Bolivia is intended to be decentralization. In April 1994 the Popular Participation Law (*Ley de Participactión Popular*, PPL) was enacted by the administration of former President G. Sanchez de Lozada. This law represented one of the largest changes in decentralization in the history of Bolivia. It led to the creation of 314 municipalities throughout the country, based on the territories of existing sections of a province as the basic administrative unit for public investment. The PPL awarded a high degree of autonomy to municipalities and radically changed investment priorities. Under this new regime, the administration of health and education services was transferred to municipalities.

In addition, local democracy was strengthened by legally recognizing grassroot organizations and traditional/indigenous authorities (territorially based organizations, known as *Organizaciónes Teritorial de Base*, OTBs), and the free election of municipal authorities such as mayors and municipal councils. In addition, surveillance committees (*Comités de Vigilancia*) were established to control municipal budgets and to look after the implementation of public works. In other words, the PPL has transferred decision making power to local governments so they can allocate their own funds as they see fit. It has broadened the participation of local actors in the use of such funds, and has also established control mechanisms for the correct use of such money.

A large proportion of funds (20 per cent of public expenditure) has been transferred from the central administration budget to the municipalities. In the former system, most transfers were at the departmental level (within nine departments). In 1992, just before the PPL was passed, 72 per cent of public investments were made by the central government, and only 0.4 per cent by municipal governments. By 1998, the central government was responsible for only 29.4 per cent of public investment, while municipal governments were responsible for 25 per cent (*Ministerio de Finanzas* 1999). This change has strengthened rural municipalities in relation to those in the large cities, which used to get 92 per cent of the general revenue transfer before the PPL was passed. Currently, rural provinces receive some 60 per cent of the funds thus transferred. According to the PPL, 90 per cent of the municipalities' budgets should be allocated to public investment.

In the first four years of PPL implementation (1994–8), most funds were spent on the construction or rehabilitation of basic infrastructure, such as repairing municipal buildings, schools, churches, construction of small roads, or paving main streets. In subsequent years, more funds have been allocated to productive infrastructure, such as small irrigation facilities, silos, slaughterhouses, and tractors.

To date, the results of the PPL implementation process have been mixed. On the one hand, the law has enhanced the capacity of local governments in rural areas to plan and execute projects. They have been allocated larger budgets and have been able to mobilize additional resources as well. It has improved the mechanisms of democracy by allowing citizens active participation, particularly those in rural areas.

However, it has achieved relatively little in terms of improving the incomes of the rural poor. Few of the investments in productive infrastructure have yielded significant economic returns. Success cases of sound economic agricultural projects financed by PPL funds are rare (Anderson 1999; Roca 1999).

## AGRICULTURAL EXTENSION: BACKGROUND AND STATUS

### POLICY CONTEXT

During much of the 1950s and 1960s, supported primarily by US funding, Bolivia implemented what can be termed the 'classic' research and extension scheme. This structure, which was similar to that used in many Latin American countries during that period, was based on a semi-autonomous institution where research was conducted in experimental stations and sub-stations, and extensionists had the role of delivering the message to farmers about new varieties, pesticides, or soil management practices. Extensionists, besides being technical assistants, also performed other roles such as leadership training, promoting organizations, helping farmers to plan, and supporting other activities beyond those of assisting in improving crops or livestock yields. This model is usually referred to in Bolivia as the Inter-American Agricultural Service (*Servicio de Agricultura Interamericano*, SAI) period. It ended in the mid-1960s when public funding for agricultural research and extension services dropped. Most of the best-qualified researchers left and infrastructure began to deteriorate (Godoy *et al.* 1993).

In 1975, the Bolivian Institute of Agricultural Technology (*Instituto Boliviano de Tecnología Agropecuaria*, IBTA) was established to provide extension and research services for the whole of Bolivia, except the Santa Cruz region, where a different and independent agricultural research centre, the *Centro de Agricola Tropical* (CIAT)[2] was created. The creation of these two centres, albeit with limited funding, boosted agricultural research and extension in Bolivia. Their methods of working and approaches were quite similar to those of the SAI period but their operations were mainly conducted by junior researchers and extensionists placed in key farming areas.

In 1991, a US$ 21 million World Bank project was approved, which aimed to strengthen IBTA's research capacity, but little of the funding was allocated to extension. In fact, all of IBTA's extension offices were closed and their personnel dismissed. The new scheme was based on the idea of producing a steady supply of technologies and assuming that farmers would search for and adopt them. Results were not as expected. The poor adoption rates of improved practices and a lack of public funding precipitated the final closure of IBTA in 1997. The only IBTA commodity programme that remained strong in terms of technology generation and extension was that of potatoes (Foundation for Research and Promotion of Andean Crops, *Fundación Boliviana para la Investigación y Promoción de los Cultivos Andinos*, PROINPA). Together with CIAT, today this foundation leads extension efforts in the country by trying new methods and approaches.

To disseminate its technologies, PROINPA has formed local agricultural research committees with farmers to conduct on-farm and participatory evaluation trials, and farmers' tours (horizontal extension). It also produces leaflets and technical bulletins, and provides video shows and training for farmers. The foundation's social and economic impact is regarded as one of the greatest in Bolivia (Fundación PROINPA 2000).

Since the closure of IBTA in 1997, the Ministry of agriculture, livestock, and rural development has been pushing a new agricultural research and extension model, the Bolivian agricultural technology system (*Sistema Boliviano de Tecnología Agrícola*). This is based on four foundations for the development of agricultural technology (Fundación para el Desarollo de Tecnología Agrícola, FDTA): one for each of Bolivia's main agro-ecological regions: the highlands, valleys, tropical semi-arid region (Chacos), and humid tropics.

Under this new model, producers are expected to demand technologies based on the principle of free market mechanisms (bidding). Agricultural research centres are also expected to compete in the bidding process through research and technology transfer projects submitted to the FDTA. The project envisions that the private sector will participate extensively, by placing demands and financing research and extension projects (IDB 1998).

INNOVATIVE EXTENSION: THE CIAT
INTERMEDIATE USERS MODEL

The CIAT is responsible for the generation and transfer of technology for the Santa Cruz region (a department of some 370,000 sq. km); it is a para-statal organization involving both public and private participation. Most of its funding comes from government regional development funds, and it has enjoyed more independence from political interference than most other para-statal organizations in the country. It has also benefited considerably from international aid.

After closing CIAT's extension department and the CDR's extension centre, CIAT executives, launched the intermediate users (IU) extension model. Agricultural research has been decentralized from the main experimental extension into local experimental units (*Centro Regional de Investigación*) in order to better replicate and be closer to the conditions of farmers in different ecological regions of Santa Cruz.

The IU approach draws on the informal contacts that CIAT researchers have formed with extensionists of non-governmental organizations (NGOs), producers associations, and agricultural inputs suppliers, to establish a formal two-way technical information flow with the producers *via* these IUs. The first task was to establish linkages with the IUs and to strengthen the publication capacity of the existing communication unit. The main mechanisms devised to formalize linkages included: (i) periodic seminars; (ii) a circulated list of technical publications; (iii) field days in the local experimental units, promoted and co ordinated with IUs; (iv) communications training for IUs (for example, how to give technical talks); and (v) participation of IUs in CIAT planning events.

In the early 1990s, the World Bank launched a lowland project that included a component to strengthen CIAT research and extension capacity. This project also built on the IU model, as it adopted a working structure based on three roles: (i) linking subject matter researchers with farmers, producers, and technical advisors (by commodity); (ii) regional co-ordination, implying the building of links between researchers of different disciplines and the IUs working in specific regional settings; and (iii) production and dissemination of technical information through various means of communication. In addition, an increased effort was made to bring researchers out of the central experimental extension unit and to conduct more trials in the local experimental units and on farmers' fields.

Thiele *et al.* (1998) identified four phases connected with dominant functions in the evolution of the extension unit within CIAT, after which the IU model was adopted: (i) bridging; (ii) adjustor (joint trials with IUs and farmers, zonal meetings, building databases); (iii) training for extensionists and forming zonal committees; and (iv) facilitator (feedback from farmers, on-field trials, discussing recommendations, guides). They stressed that the IU model increased

CIAT's senitivity to farmers' technology demands; that it has markedly increased the availability and accessibility of technologies generated by CIAT to IUs; and that operational linkage mechanisms should be tailored according to the specific functions of an extension unit.

In recent years (1999–2001), the emphasis of CIAT–DTC's[3] strategy has been oriented to increasing farmers' participation in the planning, design, execution, and evaluation of on-farm trials, while maintaining its communications and co-ordinating role with the IUs. However, due to the ending of the World Bank lowland project, the number of subject matter specialists was reduced, as was funding for extensionist training. The DTC has, therefore, tended to concentrate on a few municipalities, by applying the participatory and action approach that represents a renewed effort to improve responsiveness of the centre to farmers' demands.

The DTC selected a sample of municipalities (El Torno, Villa Florida, and Mairana) with which to conduct participatory approaches to technology transfer. In these municipalities, rapid rural appraisals were carried out. These involved considerable interaction between researchers based in the central experimental station, technical assistants, and the farmers themselves. After an analysis of the data, a list of priorities was elucidated and formed the basis for discussion of the experiments that need to be established.

The second step taken was to form community-based agricultural research centres (*Centros de Investigación Agricola Communal*, CIACs), using the same approach as PROINPA. The main duties of these centres are to: decide on the variables to be tested; help researchers to set the experiments; and record data and evaluate the results along with the researchers.

Under this scheme several experiments have been conducted on farmers' fields, for example, on new potato varieties, organic and biological pest control, silage methods, new pastures, vegetable growing, small livestock, and farming practices for maize. Throughout this process special attention has been paid to the participation of women and respect of cultural values.

The CIAT expects that as a result of these opportunities the farmers will themselves show the results to other farmers and will informally communicate the recommendations. To-date, the farmers in the sample municipalities have been enthusiastic and have shown a strong commitment to continue working under this scheme. The

CIAT has increasingly been working with some of the recently established municipalities, connecting its resources to replicate the examples in the municipality of Mairana, where the farmers' research committee submitted a project to the municipal council for counterpart funds that was accepted by the council but unfortunately not by the external donor. However, this example represents an innovative way of linking the experience of the farmer research committees with the new institutional decentralization structures.

This approach of working within the structures of municipalities is still unproven, either because most municipalities are not organized in such a way as to deal with agricultural extension, or because resource limitations prevent CIAT from working with many municipalities. Despite the enactment of the PPL in 1994, on the whole, there has been little connection with municipalities, and most of the work is conducted directly with groups of farmers.

## MAKING AGRICULTURAL EXTENSION AND RURAL DEVELOPMENT PRO-POOR: OPPORTUNITIES AND CONSTRAINTS

### MUNICIPALIZATION AND AGRICULTURAL EXTENSION

Considering that a large proportion of any municipality's income is spent on non-productive activities, the National secretariat of popular participation and several NGOs have been encouraging municipalities to shift their expenditure priorities more towards agriculture or other productive sectors such as small enterprizes. Results have been modest and have shown little impact in terms of increased productive skills.

The PPL provides municipalities with larger budgets, and depends on them to allocate funds to the most urgent or important needs. Experience has shown that most of the funds transferred by the central government have been invested in physical infrastructure and public works, and that very little has been allocated to accruing agricultural knowledge, or to improving extension services. There has been negligible investment in agricultural extension services (Ministerio de Finanzas 1999), particularly considering the small budgets that municipalities have to manage and the high demands of alternative activities. Little money has gone either to technical assistance or training. At present, very few municipalities have an agricultural department to assist farmers with their technical problems. One of the few exceptions is the Municipality of Yapacany in Santa Cruz, where one agricultural technician has been hired to co-ordinate technical assistance activities among NGOs working in the municipality. In most cases, when agricultural professionals have been hired they are given the responsibility of writing up project proposals to tap financial resources or administrative tasks, the equivalent of being independent fund managers.

What is even more worrying is that the issue is hardly ranked among the priorities of municipal development plans (Promoción al Desarrollo Economico Rural) (PADER, Cosude 1999). This can largely be attributed to a lack of clear orientation or human resources and the ambiguity of municipal functions (PADER Cosude 1999). The PPL did create some conditions conducive to improvement of the functioning of agricultural extension services at the municipal level, but for the reasons mentioned above they have failed to materialize. The decentralization, or 'municipalization',[4] of technical assistance through broad extension services is one of the missing mechanisms in the new rural development efforts.

This activity is still regarded by mayors or municipal authorities as a central/meso-government responsibility. In addition, nobody has assumed responsibility for promoting the importance of the issue or the need to develop appropriate schemes to deal with extension services. By closing down the extension service at a national level, technology transfer activities have been reduced to a minimum, and the municipalization process has not been able to replace such functions. Currently, extension services are confined to a few NGOs and to the limited activities of the DTC or CIAT in Santa Cruz. Apart from a few exceptions, municipalities have not developed an alternative scheme for two reasons. First, financial resources are limited and compete with several other important needs (health, education, roads, and sanitation). Second, physical infrastructure is more visible for political aims, easy to administer, and complies with the regulation that only 15 per cent can be spent on running costs.

Current patterns of spending, therefore, appear to be having a negligible economic impact on poor peoples' livelihoods and could perhaps be better redirected to other activities. However, the lack of investment in knowledge systems and budget orientation is not expected to change in the near future. There are no indications that the political realities that discourage such investment will change.

In Bolivia, there are no major regulations to force municipalities to provide extension services: everything depends on the mayors or municipal authorities. This can be regarded as a constraint to extension activities, particularly in the poorest municipalities. Another important constraint is the lack of knowledge, not just of the importance of extension but of its methods, approaches, and linkages with the sources of 'know how'. Again, any initiative is conditioned by the leadership (or lack of leadership) of the mayor or influential local authorities.

The great potential of PPL for extension is that it allows and induces participation. Organizations formed under the framework of this law (surveillance committees, OTBs, and municipal councils) can be used as an important forum for exchange of experiences, to build linkages with sources of technical knowledge, or as lobbying groups for such purposes. However, the catalysts that could help realize the potential of existing institutional arrangements are missing.

The failure to establish extension units (or agricultural departments) in Bolivian municipalities suggests that in the short term, at least, it is quite unlikely that municipalities will strengthen their capacity to provide this service although some of them might improve their co-ordinating role with NGOs or other intermediate users working the area. The critical factor is the lack of state initiative to regulate and promote pro-poor extension activities. There is a need for state organizations as well as NGOs to be more proactive and to re-establish the importance of the issue in the broader context of municipal and productive development.

There are only a few Bolivian examples of pro-poor extension schemes. The PPL has so far failed to provide this service in a nation-wide context; however, it does provide an important platform that could be built upon for this purpose if the catalytic policy mechanisms are instituted and activated.

IMPLICATIONS AND LESSONS FROM THE CIAT EXPERIENCE

Four extension models are interwoven in the CIAT–Santa Cruz experience:

1. the conventional institute with researchers and extensionists;
2. co-ordination between researchers and intermediate users;
3. training and making use of mass communication; and
4. participatory research involving farmers.

Models (2) and (3) have had a strong demand pull component and a relationship that is both low-cost and far-reaching from which several lessons can be learned and applied to other circumstances. Improving linkages with other extension organizations is an appropriate way of coping with limited resources and synergizing efforts. The main pitfall is the relative lack of contact with producers themselves that can lead to misinterpretation of their needs and strategies. Training intermediate users and producing extension publications is an inexpensive way of reaching the farmers, but it fails to establish the necessary contact with the producers that would enable them to orient research towards best-bet opportunities for making technical and socio-economic breakthroughs. Finally, with model (4), CIAT made a greater effort to 'listen' to the farmers' demands and certainly this is an approach that deserves credit. However, CIAT confronts the difficult trade-off between an intensive approach, essentially focusing on a few farmers (at a relative high per capita cost) and the extensive, far-reaching, low-cost approach with little contact. This dilemma has yet to be resolved, and there is a need for a more comprehensive approach that can be more demand-oriented but is within the financial possibilities of a developing country research organization.

## Endnotes

1. The HIPIC II is a G7 countries' initiative designed to alleviate the debt of developing countries by using the funds in poverty reduction programmes.

2. In this document, unless specifically referred to, CIAT refers to the Bolivian agricultural research centre for the lowlands, and not to the CIAT based in Cali, Colombia, the International Tropical Agricultural Centre (Centro Internacional de Agricultura Tropical), of the Consultative Group on International Agricultural Research (CGIAR).

3. Department of transfer and communication (*Departamento de Transferencia y Comunicación*, DTC).

4. In Bolivia, 'municipalization' is understood as the process of strengthening the functions of a municipality to conduct development efforts.

## References

Anderson, V. (1999), 'Popular participation in Bolivia: Does the law secure participation of the rural population?', CDR Working Paper 99.6, Denmark: Centre for Development Research (CDR).

Bojanic, A. and K. Hussein (2000), 'Application of guidelines for rural development profile and strategic framework of EU to the case of Bolivia', London: Overseas Development Institute—Aide à la Décision Economique, Economic Decision Support (ADE) (Belgium).

Godoy R.; M. Franco, and R. Echeverria (1993), 'A Brief History of Agricultural Research in Bolivia: Potatoes, Maize, Soybeans and Wheat Compared', Harvard Institute for International Development (HID) Discussion Paper 460, Massachusetts: HID, Harvard University.

Fundación PROINPA (2000), 'Informe de actividades', Cochabamba, Bolivia (mimeo).

IDB (Inter-American Development Bank) (1998), 'Programa de Servicios Agropecuarios BO-0176', www.iadb.org/exr/doc98

Ministerio de Agricultura, Ganadería y Desarrollo Rural (MADGR) (1999), 'Politica Nacional de Desarrollo Agropecuario y Rural', La Paz: MAGDR.

Ministerio de Finanzas (1999), 'Informe economico de la Nacion', La Paz, Bolivia (mimeo).

Montes de Oca (1997), 'Geografia y recursos naturales de Bolivia', La Paz: Editorial Don Bosco.

Morales, R. (2000), 'Bolivia: Politica Economica, Geografia y Pobreza', La Paz: Universidad Andina Simon Bolivar.

Pacheco (1998), 'Estilos de Desarrollo, deforestacion y degradacion de los bosques en las tierras bajas de Bolivia', CIFOR, CEDLA, TIERRA, La Paz: Centro de Información para el Desarrollo (CID) Pub.

PADER (Promoción al Desarrollo economico rural)/Cosude (1999), 'Municipio productivo: Aprendiendo la realidad municipal', CID pub.

PROINPA (2000), 'Promoción e investigación de productivos andinos', Cochabamba Bolivia.

Roca, C. (1999), 'Diagnostico de la Provincia Sara-Ichilo', GTZ-Prefectura del Departamento, Santa Cruz.

Thiele, G., J. Wadsworth, and R. Velez (1998), 'Creating Linkages: Lessons from Agricultural Research and Extension Liaison in Lowland Bolivia', European Journal of Agricultural Education and Extension, Vol. 4, No. 4, pp. 213-23.

World Bank (2002), 2002 World Development Indicators, CD Rom, Washington, DC: World Bank.

## Further Reading

CIAT (Centro Internacional de Agricultura Tropical) (2000), '25 Anos de Desafios y Realizaciones de CIAT', Santa Cruz: Centro Internacional de Agricultura Tropical.

———— (1999), 'Informe Tecnico 98/99', Santa Cruz: CIAT.

Republica de Bolivia (1998), 'Plan General de Desarrollo Economic y Social 1997–2002', La Paz: Presidencia de la Republica.

Velazco, M. R., G. Thiele, and P. Davies (1988), 'Estudio de la Transferencia de Tecnologia en Santa Cruz', Vol. I, Documento de Trabajo 68, Santa Cruz: MBAT—Centro Internacional de Agricultura Tropical (CIAT).

# 7
# Conclusions
## Extension and the Scope for Pro-poor Policy Reform
*Ian Christoplos • John Farrington*

The preceding chapters have been concerned with empirically documenting and analysing the 'who?' 'what?' and 'how?' of pro-poor extension. We began by asking whether extension had anything to offer to the rural poor. The cases have shown that it does, but that in order to do so extension thinking must encompass far broader conceptualizations of who the poor are, what might be considered an 'extension service', and how extension and related rural services can become far better anchored in other pro-poor policies and institutional reforms, especially decentralization. Pro-poor extension involves relating policies and reforms to the context of globalization, and the changing rural futures and extension roles that are emerging in its wake. As such it implies addressing a number of difficult issues and choices, notably:

1. Should public expenditure on extension be prioritized with an open acceptance of triage, by focusing on areas that have some opportunity of benefiting from globalization? Or can demand-based alternatives be devised for the weaker areas?

2. In a similar acceptance of triage, should extension be linked more closely with production methods focused on meeting changing consumer demands, such as selling to supermarkets, even if poor, own-account producers are left out? Might such a focus at least impact on the poorest as consumers and labourers?

3. Can a new cadre of advisors be developed that are skilled at

helping poor people make good exits from agriculture by making sound choices about their livelihoods? What type of institutions might provide a base for such 'livelihood extension'?

4. What is the potential for extension to make a more concerted contribution to addressing systemic crisis and collapse, either through information services or through closer links with public works and other safety nets? What are the implications of this in the context of weak or non-legitimate governance? What are the implications of not entering this fray?

This conclusion summarizes where we have come and where we must go in reconceptualizing the who, what, and how of pro-poor extension.

The studies have shown that, despite some notable efforts to reform, extension policy in many countries remains exclusively production-focused, institutionally monolithic, centrally directed, and organized on the premise that public sector extension structures can effectively reach down to the village level. In some countries, neo-liberal policies have led to the wide-scale privatization of extension and removal of the state 'subsidy' that it implies. In order to design extension policies and strategies that reach the poor in the development context of the twenty-first century it is important to go beyond these standard approaches. There are several reasons that a reassessment is essential if extension is to find a new role in poverty reduction.

The public sector alone cannot finance, let alone deliver, extension services to meet all requirements. Extension is being increasingly privatized in areas well-integrated into markets. This may result in resources being shifted to more remote areas where the majority of rural poor live. Some states (Vietnam and India) have acknowledged that in these remote areas, 'subsidies' in the form of publicly financed extension and other support to agricultural production are justified— they may help to prevent forms of social misery and disruption which ultimately have higher financial and human costs.

This does not, however, suggest that extension should continue along old paths even in these areas. Public sector extension is less and less able to operate effectively at the village level in remote areas, suggesting the need for a reassessment of how to prioritize the placement of public sector human resources *vis-à-vis* other extension related actors, and of how to combine public funding with private delivery. Extension provision through public sector, village-level extension workers is facing chronic difficulties. Alternatives may be found by reconsidering the role of district towns, where specialist advisors can be located in 'one-stop shops', supported by (publicly available) internet access and other information communication technologies, linking with private sector agents (technical advisory services, input supply, marketing, and processing), and adequately backed by agricultural research. Effective mobilization of the potential of pluralistic extension demands increased attention to the vertical integration of public sector extension with other actors promoting technological change. This is unlikely to happen if human resources are concentrated at the village level.

Further reforms are needed to ensure that these new forms of services are responsive to the rural poor. It is no longer adequate for extension agents to merely be accountable upwards within a line department. 'Good governance' suggests that procedures be changed to take into account the views of clients on the performance of extension workers, and to stimulate their demands on the system. Downward accountability is essential for reorienting extension to address the varied needs and priorities of the poor and to focus better on contributing to their livelihood choices. The need for this in public sector reform has long been realized. It is now being increasingly acknowledged that the private sector may not be a panacea. Where extension is publicly funded or dependent on aid flows (for example, Nicaragua, Uganda) the incentives facing extension agents in the field for upward and downward accountability are complex and require closer monitoring and deeper analysis than has often been the case to-date.

The most reliable way of ensuring that such a shift is made is by strengthening the 'demand' side of extension in ways that are cognisant of the inherent difficulties in helping the poor to make these demands. Solutions may be found in the form of 'community development' type efforts involving community groups and NGOs to identify 'para-extensionists' who will travel to district towns to articulate local requirements, bring back advice and also possibly act as commission agents for input supply and output sales. In addition, different private sector and civil society initiatives can and should fill some of the gaps emerging after the withdrawal of the state, but even these types of efforts will be limited where market integration is weak and civil institutions have been weakened by chronic instability. Elite domination of local institutions is also a danger, which must be taken into account in the promotion of participatory methods.

The spatial aspects of poverty, including changing rural–urban relations, marginalization of inaccessible hinterlands and varying impacts of globalization on local markets have created new challenges and opportunities for pro-poor extension. Pro-poor extension must have objectives of pro-poor growth as well as vulnerability reduction. Strategies emphasizing pro-poor growth will be most effective in the relatively accessible areas which are able to draw significant advantages from globalization. Vulnerability reduction is in many cases the most realistic objective for marginal areas. These categories, however, are not fixed. Even in well-integrated areas there are significant hazards from inappropriate rural development trajectories, and there is some potential for pro-poor growth even in marginal areas.

Finally, pro-poor extension is unlikely to have broad impact if it is not designed to reflect the fact that the poor are not merely producers but also labourers and consumers. Extension should pro-mote technologies that not only increase production, but also create more labour opportunities and cheaper staples for poor consumers. It should do so in ways which increase the 'voice' of poor people in their efforts to pursue a variety of livelihood strategies, and not be based on 'yeoman farmer fallacies' that assume that all those working on the land perceive own-account farming to be their best bet for survival and asset accumulation.

## 'WHO': UNDERSTANDING POVERTY

This aspect was covered in some detail in the introduction. Poor people struggle to meet their livelihood needs in very different natural environments, political structures, and market contexts. They have a range of livelihood strategies—not just own-account production. Extension that promotes technical change in agriculture must take careful account of the broader impact of technological change on the rural poor and not just the adoption of technologies by individual farmers. In order to understand the context in which pro-poor extension services are being delivered, it is important to first look at the issue of exclusion—are the poor 'integrated' into the market, the state and civil society? Do, for example, markets for goods and services function where they live? What is the government's vision regarding whether the state or private agricultural service providers should actually serve the poor? What is the rhetoric and what is the reality regarding decentralization, and how does this affect the inclusion of the poor in local decision making? What arrangements exist for ensuring the accountability of the civil service (or even private service providers contracted with public funds) and seeing to it that the poor have a 'voice' at different levels? Do the poor participate in producer and other civil society organizations?

The case studies in this volume have noted that service provision for an increasing proportion of the rural poor is being effectively abandoned due to the assumption that their livelihoods are not 'viable'. Even if one rejects such prioritization on normative grounds, if alternatives are to be developed, it is important to first understand why these exclusionary policies are emerging. This relates to the context of the rural space in terms of remoteness, proneness to natural disaster, diseases such as human immunodeficiency virus/acquired immune deficiency syndrome (HIV/AIDS), or chronic violence and political uncertainty. Weak government commitment to address the problems of marginal areas is often a pragmatic response to the limited availability of human resources and investment in places where growth is expected to be slow at best. The government may not be ready to provide subsidies indefinitely where there is little early prospect that the private sector will arrive to take over as the economic motor. Support to out-migration offers an alternative in some settings. However, this should not be migration forced by neglect, but, rather, carefully supported to prevent negative effects, and monitored at both 'origin' and 'destination' ends, to reduce the risks of saturation of job markets and exploitation of workers.

LIVELIHOOD STRATEGIES AND TECHNOLOGICAL CHANGE

Individuals are often at the same time producers, consumers, and hired labourers. Pro-poor technological change cannot, therefore, simply be geared to increased production, but must be sensitive to potential trade-offs between productivity (especially labour productivity), increased employment opportunities, and reduced vulnerability. If policies remain focused only on production itself, impacts on the poor will be slow to trickle down. There are a wide range of different poor groups in the rural space, distinguished by their development context and livelihood strategy. The provision of appropriate technologies to one poor group—low-income farmers, for example—

can have significant knock-on effects on other poor groups, and these must be taken into consideration. Technical change may displace labour where there is little prospect that those leaving agriculture will find comparable employment (or, in the case of family labour, where overall family income and well-being is diminished as a result). Wages may increase among one poor group at the expense of accompanying large-scale displacement of labour, which is then unable to find alternative work. Of particular importance is the role of induced technical change where production maximization is not the predominant livelihood objective. Households often see better prospects in reducing labour inputs on the family farm in order to more actively pursue wage labour opportunities.

Internationally and locally, globalization exerts massive influence on markets for agricultural commodities, inputs, capital, and labour. Trade theory predicts that, since developing countries have a relative abundance of unskilled labour, freer trade should increase the demand for exports that embody large inputs of unskilled labour, thus increasing employment, raising wages, and reducing poverty. However, Killick (2000) argues that increased world demand is most buoyant for products embodying skilled labour and relatively advanced technology. Globalization may, therefore, have little positive impact on the poor in developing countries, and possibly a number of negative ones. In short, globalization is creating different threats and opportunities in different contexts, which can be loosely dichotomized into areas that have relatively good access to markets for goods and labour, and those where market penetration is limited due to poor infrastructure, weak governance, and other factors (such as insecurity).

Commercialization, product processing, and marketing exemplify the former. The rapid penetration of supermarkets in developing countries presents a particular threat to small scale producers (Berdegué et al. 2000). One reason for this is their capacity to switch the sourcing of products from one country to another (even from one continent to another) in response to small, and often short-term changes in wholesale prices. This can have profoundly destabilizing implications for producers.

A second reason is rooted in growing consumer expectations (even within developing countries, and partly in response to advertising by supermarkets). These, for instance, are becoming increasingly conditioned towards cosmetic values such as uniformity in product appearance. Important preconditions here include the existence of effective processing facilities, strong organization (either from the private sector or by farmer organizations) and market information to ensure that production timing matches consumer demands and avoids flooding of urban markets. The inability of small-scale producers to control their natural environment (for example, through irrigation) or to influence the marketing and processing chain makes their chances of meeting cosmetic, timing, and quality control standards almost negligible.

A third set of reasons is rooted in health and safety considerations: the growing risk of consumer litigation is one factor underpinning pressure towards tighter health and safety requirements, especially in relation to livestock products. Some requirements (example, refrigeration of dairy products) have already impacted negatively on producers in more remote areas who face exceptionally high costs in maintaining long cold-chains,[1] even though consumer demand for cheaper and traditional dairy products has meant that some markets for unpasteurized products have proven surprisingly resilient, as in Central America. Other requirements (as for organic produce) may work in favour of producers who rely little on purchased inputs, but much of the evidence for any favourable impacts here relates to niche products, whereas mainstream products appear more likely to suffer the disadvantages (from the small-scale producer's viewpoint) of nearly unattainable quality standards and market switching.

## 'WHAT': COMPONENTS OF PRO-POOR EXTENSION

In terms of general orientation, one of the most fundamental requirements for making extension more relevant to the poor is to move it towards livelihood approaches and away from simple commodity foci, and to ensure that extension strategies embrace institution building (Box 7.1; Table 7.1). This, however, raises a fundamental dilemma, given that these activities require financial resources well beyond the levels that the state can provide, and experience and skills unfamiliar to many state employees. Again, sound understanding of the local context will be necessary before appropriate approaches can be identified.

All governments are resource constrained to varying degrees, as reflected in the steady decline of village-based extension service structures. Therefore, it is unrealistic to expect that all the elements

Box 7.1

Extension Policy: Making the Shift from a
Commodity Focus to a Livelihoods Approach

A livelihoods approach to extension implies:

- differentiating between extension for production, where a failure to directly reach the poorest is inevitable and must be accepted, and extension for livelihoods, where a broader menu of options for support should be explored;

- specifically analysing the mix of signals and incentives that are sent to frontline extension staff (objectives such as poverty alleviation, modernization, and aggregate production increase) and how the staff interpret their roles and priorities accordingly;

- critical reflection on whether expediency has led to a concentration of resources on better-off areas, despite public policies to the contrary, and if so, what this implies for poverty reduction objectives;

- ongoing analysis of what actually occurs in the wake of decentralization in order to close the gap between the rhetoric of decentralization and the often weak commitment to providing services to marginal communities and poor performance with regard to sustainability;

- designing decentralization efforts within a frank and realistic assessment of the pre-existing incentive structures, resource flows, and legitimacy of local government, civil service, and civil society institutions;

- realizing that the motor for vertical integration may not come from extension to producers alone, but rather from other actors in the commodity chain;

- analysing and supporting the information, advisory, and training needs of a wider range of actors in and around the agricultural sector, rather than only producers.

of a desirable agricultural policy or extension system can be put into place. This places a premium on careful selection of priorities, sequencing, and identification of funding sources for individual elements of a coherent strategy. For the many countries which rely heavily on aid, donor support will be required to ensure that these priorities are reflected in the new architecture of aid—for instance in poverty reduction strategy papers (PRSP) and sector-wide programmes—and that external resources are adequately sequenced and co-ordinated.

TABLE 7.1

Components of Pro-poor Extension

|  | Domains | | |
|---|---|---|---|
|  | Production | Production context | Wider livelihoods |
| Intangible factors with a potential role for extension | • Crop and livestock husbandry<br>• Farm management | • Use of natural resources in relation to agriculture<br>• Wider environmental issues<br>• Marketing<br>• Processing<br>• Producer organizations | • Alternative livelihood opportunities<br>• Improving access to natural resources for the poor<br>• Social safety nets<br>• Power relations and institutional landscape<br>• Law<br>• Local and national level planning of agricultural and natural resources (NR) 'futures'<br>• Reduced vulnerability in relation to the above |
| Institutional support with a potential role for extension | • Promotion of farmers' organizations or co-operatives for input acquisition, information access, veterinary services, phytosanitary services etc. | • Promotion of groups for joint action in NR management<br>• Promotion of farmers' organizations for marketing, processing, certification, etc. | • Promotion of groups to exercise small farm 'voice', in planning NR 'futures', in accessing NR, and in accessing alternative livelihood options and the requisite skills |
| Tangible factors with a potential role for extension | • Seeds<br>• Agrochemicals<br>• Irrigation water<br>• Productive credit | • In relation to above | • In relation to above<br>• Insurance schemes<br>• Consumption credit |

Source: Authors.

232

Even where policies are agreed upon and resources exist on the ground for their implementation, there will always be gaps between intended and actual patterns of implementation. This may be attributable variously to inadequacies in skills, to rigid divisions between remuneration and operational funds, or to rent-seeking and non-legitimate political interference. It may also be due to subsidiarity, as local actors discover the inadequacy of central level policies and adapt their efforts accordingly (which often means that extension policies are given low priority). These issues are particularly important in decentralization efforts. Potential safeguards against both poor and inflexible implementation include not relying on a single delivery agency but on a multiplicity of different kinds, preferably with some degree of competition among them when they address private goods, or co-operation in respect of public goods. Enhancing the capability of clients to make demands on the system will be important in all cases, from policy formation to implementation, and above all in the relationships between poor farmers and extension providers.

CREATING AND SUPPORTING OPPORTUNITIES FOR THE POOR

Extension can contribute to creating and supporting opportunities for the poor through increased production/productivity and access to employment. Employment opportunities can be enhanced through the promotion of labour absorbing technologies for small, medium, and large farms. Institutions can be developed that are capable, in some degree, of redressing market imperfections, including producer organizations, non-governmental organizations (NGO)/public sector models, para-extension workers, one-stop shops, and information communication technology (ICT) links with advisory services and markets.

In high potential areas a primary requirement is to enhance the capacity of the poor to identify their technology-related requirements and to select appropriate sources of advice, inputs, and marketing channels. This may often also require providing support at higher levels of the commodity chain and the broader service arena (including rural financial and business development services).

Even more isolated areas may present some prospects of linking low-income producers to markets through support for niche products such as certified organic products. Publicly-funded start-up may be necessary (given that venture capital is often unavailable), but will need to be designed and implemented so as to ensure that collaboration with the private sector is encouraged and not displaced.

ENHANCING SECURITY AND REDUCING VULNERABILITY

Vulnerability reduction involves increased resilience to livelihood shocks, environmental protection, access to safety nets and better health and nutrition. It also involves setting extension priorities within an awareness of broader threats to rural development stemming from chronic conflict/violence and even systemic collapse. In countries threatened by HIV/AIDS an effective role for extension implies strong sensitivity to the changing labour economy and realistic adaptation to declining human resource capacities in extension services. The same is true where conflict or natural disasters have disrupted pre-existing labour markets. Extension is frequently pulled in to assist with short-term (or even recurrent) cash or food-for-work programmes to shore-up the labour economy. There is potential for improving impact by using extension to better place these interventions in livelihood analyses and development trajectories.

With regard to vulnerability to floods, landslides, erosion, and other hazards, stemming from environmental destruction, increasing population, climate change, and settlement patterns, market mechanisms are of little use in mobilizing appropriate technological change. Subsidisation of rural services in general and extension in particular, to promote reduction of these risks will often be necessary. Extension may even have a contribution to make in addressing chronic insecurity. Without tangible improvements in people's livelihoods in poorer regions, the cycles of conflict that are apparent in many parts of the world will only be reinforced. Rural violence can be mitigated through livelihood opportunities for youth and marginalized groups. Despite profound challenges, extension is one of the few tools to promote livelihood stabilization in areas plagued by chronic conflict.

In general, since extension structures are often drawn into post conflict and post natural disaster rehabilitation efforts, greater systemic efforts should be made to ensure that the potential of linking such programmes to goals of reducing risk and contributing to overall development are achieved.

ENABLING EMPOWERMENT

In all of these areas, the poor will need a stronger stance in dealing with institutions of government and the market if they are to transform production increases into better livelihoods. Power is related to knowledge of the market for their products, ability to update that knowledge, and institutions that create a critical mass for negotiation and a choice of production options. Furthermore, power is dependent on poor people having a choice among different livelihood options. Extension has a significant potential role to play in increasing entitlements to the knowledge and resources required to pursue a broader range of livelihoods. It can deal with some of these factors directly. In others, its role will need to be developed within a broader policy and institutional environment that enhances the power of poor people to exert their demands.

There are a number of options that can be considered for enhancing empowerment. The most obvious is local-level formation of farmer groups. These efforts must be cognisant of structural constraints to participation by the poor and the need to base such support in broader institutional structures. There may be potential for development of advisory services within producer organizations.

Empowerment may also emerge through alternative financing mechanisms for the targeted transfer of purchasing power to farm households (example, through vouchers), which they can then spend on the most appropriate source of technical innovation and expertise. Rehabilitation programmes and other safety nets may provide important windows of opportunity for this.

Decentralized systems of administration and local government may encourage responsiveness and accountability of extension services to local people. In some circumstances, it may be desirable to make the extension administration formally accountable to the local government, or even to encourage the tendency of NGOs and farmer organizations to 'contract-in' public sector extension services by paying them to provide certain services.

## RETHINKING EXTENSION AND THE ROLE OF THE STATE

New visions for extension suggest that governments should primarily act as enablers and not merely as service providers. In order for this to lead to genuine impact, extension policy must transcend the narrow focus on existing agricultural extension bureaucracies. In this sense, two recognitions, originating at opposite ends of the spectrum, are vital. First, where markets function well, a high degree of private provision of advisory services can be anticipated, leaving the state to reduce its role to enabler (either directly, or by funding others) of provision of public goods. This then frees resources for allocation to weakly integrated areas characterized by a high incidence of poverty and market failure. Second, there is growing evidence of the chronic non-sustainability of publicly-provided extension in these poorly integrated areas: it is difficult to keep staff in post and provide them with adequate back-up; fiscal crises in the state exacerbate this; and the technical complexities of new production opportunities demand higher levels of skill and back-up than can feasibly be provided (Box 7.2).

A major role for the state as enabler is to empower the rural poor to meet their own technology needs and to make effective demands on providers of extension, inputs, and marketing advice. In this context, it might pursue varying combinations of the following:

1. skills that increase producers' power to negotiate (knowledge of marketing, quality control, and certification bureaucracies);

2. infrastructure that increases producers' power to negotiate (storage, processing, and other post-harvest technologies);

3. organizations and institutions that increase producers' power to negotiate and demand services;

4. the existence of more than one person for the poor to negotiate with (that is, more competition among traders and a more dynamic service economy);

5. control of the production process through producer capacity to manage linkages of credit, processing, marketing, quality control, and input supply; and

6. diversification to avoid dependence on one crop/buyer/processing structure.

## CONSIDERING SPATIAL PRIORITIES

One of the central aspects to be considered when deciding how to promote pro-poor extension is that of where the poor actually live. This determines the potential for helping poor people to engage in market opportunities or, alternatively (especially where market opportunities are very limited), helping them cope with their vulnerabilities by reducing risk and improving subsistence production and consumption. The following scheme suggests how one might differentiate priorities in high potential and marginal areas respectively.

Box 7.2

Extension for the Poorest?

A question that must ultimately be addressed in assessing if and how extension can become more pro-poor is whether or not it can reach the poorest of the poor. Even if national policies ostensibly promote pro-poor extension, in practice services may fail to reach the extremely poor and destitute due to:

- lack of research capable of generating appropriate technologies;
- lack of logistical capacity to reach the isolated poor;
- lack of staff and/or institutional structures in poor areas;
- poor ethnic relations and lack of language skills for working with ethnic minorities;
- insecurity in marginal areas;
- poorly functioning marketing structures that obstruct input supply and market access;
- lack of knowledge and capacity within the extension organization for judging markets for non-traditional crops for which the poor may have a comparative advantage;
- absence of policy directives to explicitly take into account labour markets or consumption factors in extension priorities;
- underlying assumptions that targeting the poor is the role of NGOs and donor-financed projects, or some ministry other than agriculture.

It is clear from this list that reaching the poorest of the poor is not merely a matter of tweaking existing structures. Extension for this target group may best be developed within structures outside of regular line structures in ministries of agriculture, and in ways synergistic with resources from social protection programmes, rehabilitation projects, and civil society. As one example, where cash transfers (such as pensions) are made to those who cannot easily engage with the productive economy by, example petty trading or selling their labour, this can help to enhance and stabilize local demand for agricultural products, and so provide much more valuable support to small farmers producing staples than, for instance, food transfers would. If such 'out-of-the-box' approaches are to be related to extension policies, it is furthermore important to ask whether (i) there is a normative commitment among key stakeholders to reaching the very poorest, and (ii) whether there are viable options for reaching the very poorest where they live and within their existing livelihood strategies.

In high potential and accessible areas, the major focus should be to support the growth of both direct (own account) and indirect (wage labour) opportunities. The private sector will dominate the agenda for technological change in accessible areas. The public sector has a relatively limited role, and should emphasize clearly defined public goods, especially those related to addressing vulnerability to livelihood shocks and social protection related interventions in environmental health, sanitation, and nutrition. Labour markets should be a major factor in programming, albeit with an acceptance of the fact that government policy can influence but presumably not lead the development of new labour markets. There is also a role for the public sector to provide technical back-up to re-establish production after a disaster, where the private sector is overwhelmed, and where capital is in short supply. Possible areas of focus include:

1. commercialization of fruit, vegetable, livestock, and dairy production;

2. expanded irrigation;

3. labour-saving technologies for household production;

4. labour-intensive technologies for large-scale production;

5. environmental health interventions; and

6. improvement of production quality, timeliness, and sanitation.

Coping strategies will dominate the agenda for many low potential and isolated areas, although some openings do exist for limited growth oriented strategies. To suggest that public sector extension should buck the overall trend of retrenchment from low potential and isolated areas is rather over-optimistic. There is, however, a role for public-service institutions to participate in finding synergy between social protection, temporary safety nets and 'normal' development. Possible areas of focus in poorly integrated areas include:

1. products with high value relative to transport cost;

2. diversification of diets;

3. subsistence production;

4. natural resource and watershed management;

5. more effective use of safety nets and closer synergy between production focused and social protection strategies;

6. more effective links with relief and rehabilitation programmes through a focus on risk and vulnerability; and

7. skills and other support for migration.

The state needs to continue to play a role in relation to extension, though this will vary given the circumstances and will often require a realignment of its relationship with other actors and agencies. Where the state can have a role in the provision of extension, it may vary from being a service deliverer to being an enabler (Table 7.2). Clearly the tendency in many situations is for the former role in delivery to be reduced due to resource constraints. Here we consider what roles the state can play as an enabler of best practice in extension approaches and methods to enhance service levels for the benefit of the poor.

## 'HOW': POLICY REFORM FOR SUPPORTING PRO-POOR EXTENSION

In terms of policy design and implementation, extension programming alone cannot successfully be reoriented to address the different ways that technological change and access to technological knowledge impact the poor without a supportive reorientation of agriculture, rural development and, increasingly, urban development policy. Ideally, processes underpinning the new architecture of aid, such as the preparation of PRSPs', should support the re-orientation of extension policy by eliciting perspectives of the poor which are useable in extension design; by fostering greater coherence among extension and agricultural and rural development policy, and between donor and government priorities; and by placing extension policy within a realistic analysis of the changing social, political, and economic context of rural development, especially regarding the challenges of marginal areas.

### PRO-POOR RURAL DEVELOPMENT POLICY REFORM

The studies in this volume have clearly shown that there is no 'ideal' policy. Individual countries will have to make choices based on their specific circumstances and on the resource constraints they face. Evidence suggests that the scope for incorporating pro-poor dimensions through changes in existing policies may range from the modest to the more radical. In an ideal world, governments should work

TABLE 7.2

Policies to Support Pro-poor Extension According to Market Conditions and Characteristics of Government

| | Market conditions | |
|---|---|---|
| | Areas highly integrated into markets | Areas weakly integrated into markets |
| Government adequately resourced and responsive to needs of poor | (i) move towards fully privatized model of extension, limiting public involvement to facilitating and regulatory functions, promoting labour absorption where possible, and switching public resources to poorly integrated areas<br><br>(ii) State regulation and/or funding needed for public goods and addressing market failure<br><br>(iii) finance gaps in advisory services, particularly in making shifts from extension on production to production context, and service agency start-ups; links to commercialization and processing structures;<br><br>(iv) finance start-ups in niche enterprizes, emerging sectors, capture higher value links in commodity chains. | (a) substantial public funding generally justified<br>(b) promote public/private (commercial and non-profit) partnerships in niche products, or in adding extension to private input supply<br>(c) promote ICTs to reduce information gaps in more remote areas<br>(d) aim to raise labour productivity and release some labour for out-migration<br>(e) enhance relevance of advice to production context<br>(f) aim to reduce vulnerability and broaden safety nets (watershed management, but also crop insurance, pensions)<br>(g) co-ordinate planning at national and local levels on rural futures, example, to allow remote areas to link into commodity chains in ways reflecting their competitive advantage, including markets for niche products<br>(h) make target areas accessible by new infrastructure to ensure that poor farmers can take advantage of new opportunities<br>(i) strengthen district towns as a locus of provision of agriculture services, and people's capacity to make demands on these<br>(j) in general, keep the focus of extension broad, to embrace the wider production and livelihood contexts |

(Contd.)

240

TABLE 7.2: contd.

| | Market conditions | | |
|---|---|---|---|
| | Areas highly integrated into markets | Areas weakly integrated into markets | |

| Government adequately resourced but unresponsive to needs of poor | iii. Government policy should move towards (i) and (ii) above, but likely to be characterized by top-down service provision and by tension between unresponsive centre and more progressive local government | (j) Government policy should move towards (b) to (j), but top-down service provision likely to dominate, and to be characterized by gaps, irrelevancies, and non-sustainability in remote areas. Likely to be very limited work in partnership with other types of organization. |
|---|---|---|
| Government poorly resourced | Responsive governments will aim to move towards (i) and (ii) and (b) to (j), but will be severely resource constrained, and will have to rely on partnership or 'intermediate user' arrangements. Less responsive governments will impose a *de facto* triage, ignoring areas that are beyond their effective control. If poorly managed, this may lead to 'durable disorder', as alienation and chronic violence take hold in the hinterlands. | |

*Source:* Authors.

241

towards designing and implementing extension policies that take account of the following realities.

1. Poor people manage diverse portfolios, often have objectives other than increased income (such as greater well-being, increased voice, and reduced vulnerability), switch within their portfolios according to changing circumstances, and pursue their perceived rights to access a range of resources and entitlements in support of their livelihood strategies. The governments should recognize that support for the poor in some circumstances might be achieved more by interventions to reduce vulnerability than to enhance productivity, and that extension can affect the poor not merely as producers, but also as consumers or labourers.

2. The majority of the poor live in areas weakly integrated with markets; that remoteness protects them from some of the negative effects of globalization, but also—and more importantly—prevents them from accessing its benefits. Sensitive policies are needed to open opportunities for them (including the opportunity of labour out-migration), and to declare their way of life 'non-viable' increases their vulnerability and may contribute to alienation and civil strife. Policies should also recognize that globalization needs to be addressed by a combination of strategies: to highlight the negative impacts of Northern protectionism in international fora; to protect their producers where this can legitimately be done; and to design agricultural and rural development strategies that provide low income producers access to domestic and international markets in ways consistent with their competitive advantage.

3. Local government should be reinforced where it can contribute to a strengthening of local economic, political, and social life. A role for state financing (and possibly, delivery) of services will remain valid in many areas, well beyond the 'facilitating and regulating' provisions of neo-liberalism, and substantial support needs to be given to civil society organizations if they are to play their due role in making demands on the state, but also on occasion in direct provision of services such as extension. The key issue is to focus the role of the state on provision of services to the poor in relation to market failure and of public goods by targeting provision of services and entitlements to the poor.

4. Policies towards pluralism are needed in order to reassess the role of public and private sectors, identify opportunities for partnership,

and recognize the plurality of sources—nationally and beyond—from which information relevant to the rural poor can be drawn. Extension policy is not just about policies towards extension agencies. The roles of agricultural extension agencies should be defined within policies that facilitate the involvement of a far broader range of actors involved in 'livelihood extension'. This involves making a realistic assessment of the capacity of the private sector, accepting that building capacity in the private sector and addressing market failure often require positive public action and not just a 'hands-off' approach. Public delivery of services may, therefore, still have to predominate in remote areas, and mixed 'partnership' arrangements may be necessary elsewhere, other than in areas where commercial relations dominate.

5. Policies are needed that combine support to available production opportunities with mitigation of the risks associated with changing settlement and protection patterns. To declare areas that are chronically threatened by flood cyclone etc. as 'non-viable' would, in its logical conclusion, require massive (and perhaps, forced) relocation of people—in many cases with no clear alternative destination or livelihood. It also needs to be recognized that, although government may have difficulty in accessing areas of chronic political instability, some kind of production system and market may continue to operate in these areas and would merit support from those (example, private sector, NGOs) who are capable of entering them to reduce the threat of greater marginalization and systemic collapse.

6. Spatial policies should be based on awareness of where the poor are located, and what they already (or could) produce with characteristics competitive with what is produced elsewhere. Policies relevant to poorly integrated areas will be different from those relevant to favoured areas; and in all cases, close coherence will be required between national and provincial level policies if poorly integrated areas are to have some prospect of exploiting their competitive advantage. Efforts to reduce the 'poverty of place' will require both conventional and non-conventional (example, ICTs) infrastructure based on a more profound understanding of the links between urbanization (especially the role of small district towns), diversification, and commercialization. Investment may be most effective if directed towards the concentration of services into one-stop shops located in district towns.

7. Subsidies are a legitimate public policy instrument, but there is a need to target them in relation to desirable and feasible improvements in the functioning of markets, through subsidies to the development of infrastructure and institutions, in ways which support rather than crowd out nascent private activity. This implies the need to allocate a substantial part of government expenditure to the provision of public goods, including policies and actions which are spatially differentiated, as between commercial and well-integrated areas, more remote and poorly integrated areas, and those suffering systemic collapse, with a gradual shift in subsidy from the first to the other two.

8. Labour market interventions must be related to intended trajectories of technological change by analysing, acknowledging and strategizing on the indirect effects of agricultural and rural development policy on the poor, particularly in terms of labour opportunities. This requires greater coherence between the recognition of the importance of labour markets that is emerging in overall poverty alleviation goals and the production and 'modernizing' focus of many agricultural policies—by incorporating poverty analysis and monitoring into private sector investment and institution building initiatives. A spatial analysis needs to be applied to poverty and labour markets that recognizes the relationship between poverty and isolation, and also includes a focus on the situation of labourers and marginal producers who live in higher potential areas.

9. Consultative processes are important to identify what the poor see as priorities, and to obtain their views on wider possibilities which the poor may not be aware of. However, this must be moderated by pragmatic acknowledgement of the declining potential for establishing viable village-level extension structures and intensive face-to-face participatory methodologies that actually reach isolated and marginalized poor communities.

10. Rejuvenated community development initiatives are required to address poverty from a livelihoods and empowerment perspective and not merely from the individual perspectives of production or protection. These initiatives should be promoted, not just as channels for receipt of extension and related services, but also to empower producers and local organizations to better draw-down the services that they require. Realism is necessary regarding the complexity of such initiatives, as they may require the integrated provision of

services across a range of sectors, including health, education, communications and finance, as well as agriculture production and marketing oriented services.

## CONCLUSION

Extension thinking has moved beyond seeing extension actors as merely being a tool for delivering research results to farmers. While the extension discourse has become broader, it has not necessarily become clearer on what extension should and should not do. Political and policy choices will need to be made if significant levels of investment in extension services are to be maintained. If these policies demand that extension becomes genuinely pro-poor, it must be looked at in a different light with more explicit normative goals. As such it must become a broader part and parcel of rural development and poverty alleviation policies that are driven by principles of:

1. creating and supporting opportunity by looking for where the poor can benefit from the market, as producers, labourers, and consumers;

2. enhancing security and reducing vulnerability by designing strategies and priorities that avoid or reduce risk;

3. enabling empowerment by helping the poor to develop their power in relation to market actors and extension providers; and

4. institutional reform by redefining the role of the state and other institutional actors in their interactions with the poor.

The findings in this volume suggest that there is a tremendous need for pro-poor services that could, for want of a better word, be called 'extension'. There is also an array of actors that are providing and trying to provide such services. These actors often do not, however, have the job title of 'extension agent'. In Vietnam and India, efforts are underway, some more successful and some less, to reform existing agencies. In Bolivia, the former extension service has more or less disappeared, and although there are interesting initiatives afoot, their ultimate survival depends on extension becoming a priority for local government, something that has yet to happen. In Nicaragua and Uganda the bits and pieces of the former extension service are struggling to join up with newly recognized extension actors to find solutions different from those that were applied in the past. Again, the results are varied, and not always to the benefit of the poor. Decentralization creates opportunities for and threats against these efforts. In most of these cases the missing links in supporting pro-poor extension are policy visions for rural development and governance structures to ensure the accountability of frontline service providers to their clients. These problems can be addressed only in a broad-based and sustainable manner if poverty alleviation becomes a foundation, rather than an add-on, in extension thinking.

### Endnotes

1. The example of small-scale remote milk producers in Argentina cited by Berdegué and Reardon is particularly informative.

# Index